FIBER OPTICS

SCIENCE SERIES

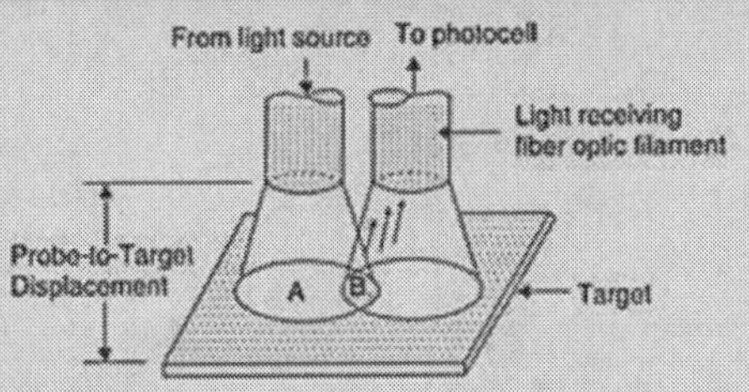

Fig. 37: Object moving away from probe, causes increase in reflected light intensity.

Construction: Fig. 37 shows the arrangement of a displacement sensor. Two separate optical fibres are positioned adjacent to each other. One of them transmits light coming from a light source. The other fibre receives light reflected from the object under study and passes it on to a photodetector.

Working: Light from the transmitting fibre element is incident on the object under study. The light receiver fibre element is positioned adjacent to the transmitting fibre. If the gap between the object and the fibre elements is zero, the light from the transmit fibre would be directly reflected back into itself and little or no light would go into the receive fibre. When the object moves away, the gap increases and some of the reflected light is captured by the receive fibre which in turn is carried to the photodetector. As the gap increases, a distance will be reached at which maximum reflected light is received by the photodetector. Further increase in the gap will result in a decrease in the light at the receiver fibre face and corresponding drop in the signal output from the photodetector. By proper calibration, we can obtain the displacement of the object in terms of the strength of the output signal of the photodetector.

23.3 FORCE SENSOR

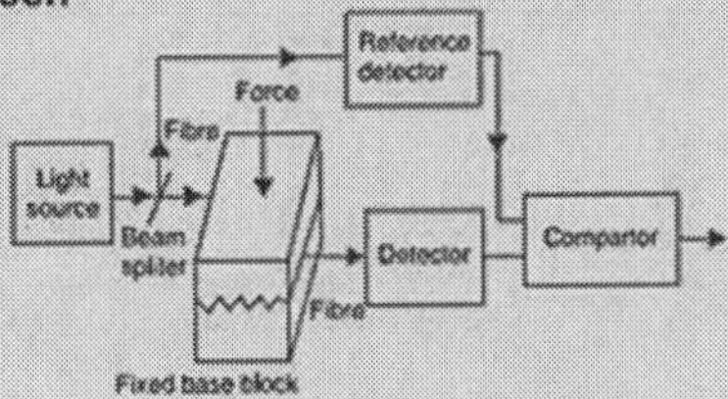

Fig. 38: Force sensor using microbend losses

Principle: This sensor is based on variations of light intensity. When an optical fibre is pressed, a small change occurs in light propagation direction due to microbending of the fibre. As a result, energy from one mode is transferred to another mode through mode coupling. In addition, higher order modes are likely to change into radiation modes. All these effects cause a loss in intensity of the light transmitted through the fibre. Therefore, the change in intensity of the transmitted light is proportional to the force applied on the optical fibre.

BY:AHMED ADEL

SCIENCE
SERIES
FIBER OPTICS
BY:AHMED ADEL

Booklet Description: "SCIENCE SERIES, FIBER OPTICS"
Explore the fascinating world of fiber optics in this comprehensive booklet, "SCIENCE SERIES, FIBER OPTICS." Spanning 46 pages, this 6 x 9 inch guide is designed to provide readers with a thorough understanding of the principles and applications of optical fibers.

Contents Overview

Introduction: An engaging overview of fiber optics, setting the stage for deeper exploration.

Optical Fibre: Detailed explanations of what optical fibers are and their significance in modern technology.

Propagation of Light Through an Optical Fibre: Insights into how light travels within these fibers, including key principles.

Fractional Refractive Index Change: A discussion of how variations in refractive index affect light propagation.

Numerical Aperture: Understanding the concept of numerical aperture and its importance in fiber optics.

Skip Distance and Number of Total Internal Reflections: Examining the conditions that govern light reflection within optical fibers.

Modes of Propagation: An overview of different modes in which light can propagate through fibers.

Types of Rays: A look at the various rays and their roles in fiber optics.

Classification of Optical Fibres: An exploration of how optical fibers are categorized based on their characteristics.

Three Types of Fibres: Detailed descriptions of the main types of optical fibers.

Materials: Insights into the materials used in the fabrication of optical fibers.
V-Numbers: Explanation of V-numbers and their relevance in fiber optics.
Fabrication: A guide to the manufacturing processes involved in creating optical fibers.
Losses in Optical Fibre: Discussing the various types of losses that can occur in fiber optics and their implications.
Distortion: Examination of how distortion affects signal quality and integrity.
Bandwidth: Understanding the bandwidth capabilities of optical fibers.
Characteristics of the Fibres Splicing: Insights into fiber splicing techniques and their importance in communication systems.
Application: A look at the wide-ranging applications of fiber optics across different fields.
Fibre Optic Communication System: An in-depth exploration of how fiber optics revolutionized communication.
Merits of Optical Fibres: Highlighting the advantages of using optical fibers over traditional methods.
Fibre Optic Sensors: An overview of the innovative uses of fiber optics in sensor technology.
This booklet serves as an essential resource for students, educators, and professionals interested in the science and technology of fiber optics, providing both foundational knowledge and advanced insights into this rapidly evolving field.

CONTENTS

1. INTRODUCTION

In 1870 John Tyndall, a British physicist demonstrated that light can be guided along the curve of a stream of water. Owing to total internal reflections light gets confined to the water stream and the stream appears luminous. A luminous water stream is the precursor of an optical fibre. In the 1950's, the transmission of images through optical fibres was realized in practice.

2 OPTICAL FIBRE

Definition: *An optical fibre is a cylindrical wave guide made of transparent dielectric, (glass or clear plastic), which guides light waves along its length by total internal reflection.* It is as thin as human hair, approximately 70 μm or 0.003 inch diameter. (Note that a thin strand of a metal is called a *wire* and a thin strand of dielectric materials is called a *fibre*).

Optical Fibre.

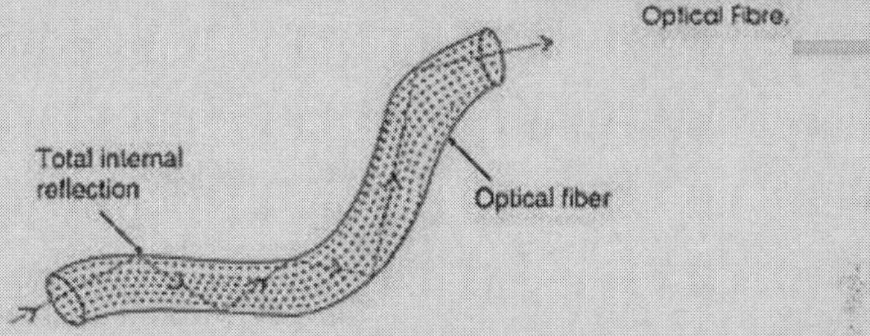

Fig. 1: Illustration of a transparent fibre guiding light along its length.

Principle: The propagation of light in an optical fibre from one of its ends to the other end is based on the principle of *total internal reflection*. When light enters one end of the fibre, it undergoes successive total internal reflections from sidewalls and travels down the length of the fibre along a zigzag path, as shown in Fig.24.1. A small fraction of light may escape through sidewalls but a major fraction emerges out from the exit end of the fibre, as illustrated in Fig. 24.1. Light can travel through fibre even if it is bent.

Structure:

Fig. 2: Side view and cross sectional view of a typical optical fibre

A practical optical fibre is cylindrical in shape (Fig. 24.2a) and has in general three coaxial regions (Fig. 24.2b).

(*i*) The innermost cylindrical region is the light guiding region known as the **core.** In general, the diameter of the core is of the order of 8.5 μm to 62.5 μm.

(*ii*) It is surrounded by a coaxial middle region known as the **cladding.** The diameter of the cladding is of the order of 125 μm. The refractive index of cladding (n_2) is always lower than that of the core (n_1). Light launched into the core and striking the core-to-cladding interface at an angle greater than critical angle will be reflected back into the core. Since the angles of incidence and reflection are equal, the light will continue to rebound and propagate through the fibre.

(*iii*) The outermost region is called the **sheath** or a **protective buffer coating**. It is a plastic coating given to the cladding for extra protection. This coating is applied during the manufacturing process to provide physical and environmental protection for the fiber. The buffer is elastic in nature and prevents abrasions. The coating can vary in size from 250 μm or 900 μm. To sum up

- Core is the inner light-carrying member.
- Cladding is the middle layer, which serves to confine the light to the core.
- Buffer coating surrounds the cladding, which protects the fibre from physical damage and environmental effects.

2.1 NECESSITY OF CLADDING

The actual fibre is very thin and light entering a bare fibre will travel along the fibre through repeated total internal reflections at the glass-air boundary. However, bare fibres are used only in certain applications. For use in communications and some other applications, the optical fibre is provided with a cladding. *The cladding maintains uniform size of the fibre, protects the walls of the fibre from chipping, and reduces the size of the cone of light that will be trapped in the fibre.*

- It is necessary that the diameter of an optical fibre remains constant throughout its length and is surrounded by the same medium. Any change in the thickness of the fibre or the medium outside the fibre (when the fibre gets wet due to moisture etc) will cause loss of light energy through the walls of the fibre.
- A very large number of reflections occur through the fibre and it is necessary that the condition for total internal reflection must be accurately met over the entire length of the fibre. If the surface of the glass fibre becomes scratched or chipped, the normal to the edge will no longer be uniform. As a result, the light traveling through the fibre will get scattered and escapes from the fibre. This also causes loss of light energy.
- Part of light energy penetrates the fibre surface. The intensity of the light decreases exponentially as we move away from the surface, as the light is able to penetrate only a very small distance outside the fibre. However, anytime the fibre touches something else, the light can leak into the new medium or be scattered away from the fibre. This effect causes a significant leakage of the light energy out of the fibre. Even a small amount of dust on the surface would cause a fair amount of leakage.
- If bare optic fibres are packed closely together in a bundle, light energy traveling through the individual fibres tends to get coupled through the phenomenon of *frustrated total internal reflection*. Cladding of sufficient thickness prevents the leakage of light energy from one fibre to the other.

  The fiber is provided with a cladding in order to prevent loss of light energy due to the above reasons.
- The cladding causes a reduction in the size of the cone of light that can be trapped in the fibre. Light entering the fibre at larger angles will strike the fibre walls at smaller angles (higher modes) and ultimately travel a longer distance. Such higher modes of a light signal will take longer time to reach the end of the fibre than the lower modes. Therefore, a pulse sent through optical fibre spreads out. The spreading would be larger, the larger the cone of acceptance. Such pulse spreading limits the rate of data transmission through the fibre. As fibers with a cladding have smaller cone of acceptance, they carry information at a much higher bit rate than those without a cladding.

Thus, the cladding performs the following important functions:

- Keeps the size of the fibre constant and reduces loss of light from the core into the surrounding air.
- Protects the fiber from physical damage and absorbing surface contaminants.
- Prevents leakage of light energy from the fibre through evanescent waves.
- Prevents leakage of light energy from the core through frustrated total internal reflection.
- Reduces the cone of acceptance and increases the rate of transmission of data.
- A solid cladding, instead of air, also makes it easier to add other protective layers over the fibre.

2.2 OPTICAL FIBRE SYSTEM

An optical fibre is used to transmit **light signals** over long distances. It is essentially a **light-transmitting medium**, its role being very much similar to a coaxial cable or wave-guide used in microwave communications. Optical fibre requires a **light source** for launching light into the fibre at its input end and a **photodetector** to receive light at its output end. As the diameter of the fibre is very small, the light source has to be dimensionally compatible with the fibre core. Light emitting diodes and laser diodes, which are very small in size, serve as the light sources. The electrical input signal is in general of digital form. It is converted into an optical signal by varying the current flowing through the light source. Hence, the intensity of the light emitted by the source is modulated with the input signal and the output will be in the form of light pulses. The light pulses constitute the signal that travels through the optical fibre. At the receiver end, semiconductor photodiodes, which are very small in size, are used for detection of these light pulses. The photodetector converts the optical signal into electrical form. Thus, a basic *optical fibre system* consists of a LED/laser diode, optical fibre cable and a semiconductor photodiode.

2.3 OPTICAL FIBRE CABLE

Optical fibre cables are designed in different ways to serve different applications. More protection is provided to the optical fibre by the "cable" which has the fibres and strength members inside an outer covering called a "jacket". We study here two typical designs:a single fibre cable or a multifibre cable.

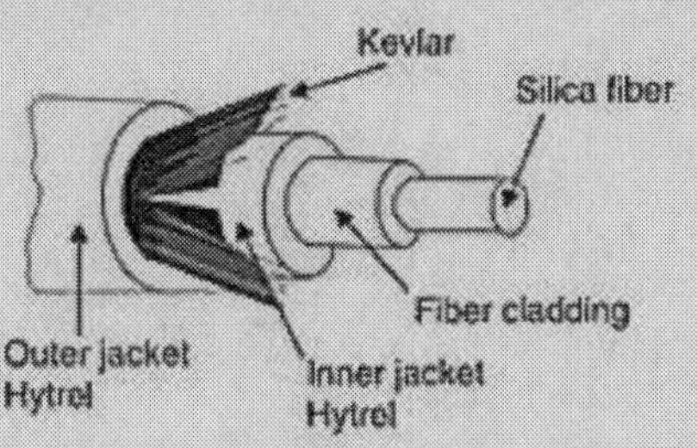

Fig. 3: Single fibre cable

- **Single Fibre Cable:** Around the fibre a tight buffer jacket of Hytrel is used (see Fig. 3). The buffer jacket protects the fibre from moisture and abrasion. A strength member is arranged around the buffer jacket in order to provide the necessary toughness and tensile strength. The strength member may be a steel wire, polymer film, nylon yarn or Kevlar yarn. Finally, the fibre cable is covered by a Hytrel outer jacket. Because of this arrangement fibre cable will not get damaged during bending, rolling, stretching or pulling and transport and installation processes. The single fibre cable is used for indoor applications.

90°, as seen in Fig. 5(b). At angles greater than θ_c there are no refracted rays at all. The rays are reflected back into the denser medium as though they encountered a specular reflecting surface (Fig. 24.5c). Thus,

- If $\theta_1 < \theta_c$, the ray refracts into the rarer medium
- If $\theta_1 = \theta_c$, the ray just grazes the interface of rarer-to-denser media
- If $\theta_1 > \theta_c$, the ray is reflected back into the denser medium.

The phenomenon in which light is totally reflected from a denser-to-rarer medium boundary is known as **total internal reflection**. The rays that experience total internal reflection obey the laws of reflection. Therefore, the critical angle can be determined from Snell's law.

When $\theta_1 = \theta_c$, $\theta_2 = 90^\circ$.

Therefore, from equ.(24.1), we get

$$\mu_1 \sin\theta_c = \mu_2 \sin 90^\circ = \mu_2$$

$$\therefore \quad \sin\theta_c = \frac{\mu_2}{\mu_1} \qquad 2$$

When the rarer medium is air, $\mu_2 = 1$ and writing $\mu_1 = \mu$, we obtain

$$\sin\theta_c = \frac{1}{\mu} \qquad (24.3)$$

4 PROPAGATION OF LIGHT THROUGH AN OPTICAL FIBRE

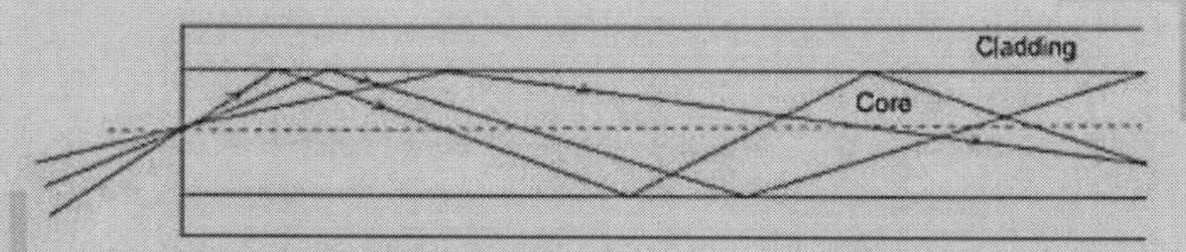

Fig. 6 : Propagation of light rays through an optical fibre due to total internal reflection.

The diameter of an optical fibre is very small and as such we cannot use bigger light sources for launching light beam into it. Light emitting diodes (LEDs) and laser diodes are the optical sources used in fibre optics. Even in case of these small sized sources, a focusing lens has to be used to concentrate the beam on to the fibre core. Light propagates as an electromagnetic wave through an optical fibre. However, light propagation through an optical fibre can as well be understood on the basis of *ray model*. According to the ray model, light rays entering the fibre strike the core-clad interface at different angles. As the refractive index of the cladding is less than that of the core, majority of the rays undergo total internal reflection at the interface and the angle of reflection is equal to the angle of incidence in each case. Due to the cylindrical symmetry in the fibre structure, the rays reflected from an interface on one side of the fibre axis will suffer total internal reflections at the interface on the opposite side also. Thus, the rays travel forward through the fibre via a series of total internal reflections and emerge out from the exit end of the fibre (Fig. 6). Since each reflection is a total internal reflection, there is no loss of light energy and light confines itself within the core during the course of propagation. Because of the negligible loss during the total internal reflections, optical fibre can carry the light waves over very long distances. Thus, the optical fibre acts essentially as a wave-guide and is often called a **light guide or light pipe.** At the exit end of the fibre, the light is received by a photo-detector.

Total internal reflection at the fibre wall can occur and light propagates down the fibre, only if the following two conditions are satisfied.

1. The refractive index of the core material, n_1, must be slightly greater than that of the cladding, n_2.
2. At the core-cladding interface (Fig 7), the angle of incidence ϕ between the ray and the normal to the interface must be greater than the critical angle ϕ_C defined by

$$\sin\phi_C = \frac{n_2}{n_1} \qquad 4$$

It is to be noted here that only those rays, that are incident at the core-clad interface at angles greater than the critical angle will propagate through the fibre. Rays that are incident at smaller angles are refracted into the cladding and are lost.

4.1 CRITICAL ANGLE OF PROPAGATION

Let us consider a step index optical fibre into which light is launched at one end. The end at which light enters the fibre is called the **launching end.** depicts the conditions at the launching end. In a step-index fibre, the refractive index changes abruptly form the core to the cladding. Now, we consider two rays entering the fibre at two different angles of incidence.

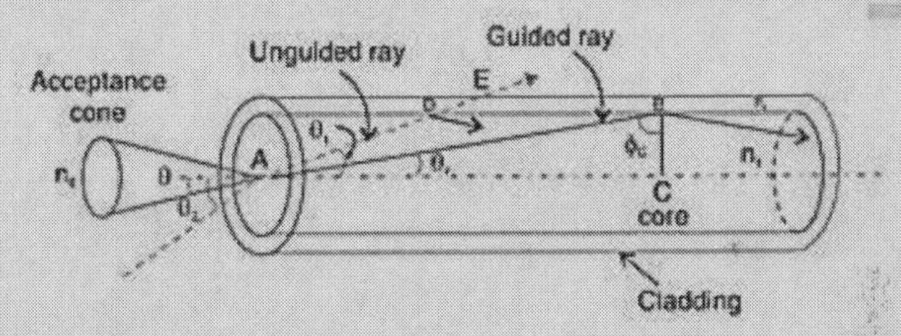

Fig. 7: Light rays incident at an angle smaller than critical propagation angle will propagate through the fibre.

The ray shown by the broken line is incident at an angle θ_2 with respect to the axis of the fibre. This ray undergoes refraction at point A on the interface between air and the core. The ray refracts into the fibre at an angle θ_1 ($\theta_1 < \theta_2$). The ray reaches the core-cladding interface at point D. At point D, refraction takes place again and the ray travels in the cladding. Finally, at point E, the ray refracts once again and emerges out of fibre into the air. It means that the ray does not propagate through the fibre.

Let us next consider the ray shown by the solid line in Fig. 7. The ray incident at an angle θ undergoes refraction at point A on the interface and propagates at an angle θ_c in the fibre. At point B on the core-cladding interface, the ray undergoes total internal reflection, since $n_1 > n_2$. Let us assume that the angle of incidence at the core-cladding interface is the *critical angle* ϕ_c where ϕ_c is given by

$$\phi_C = \sin^{-1}(n_2 / n_1) \qquad 4a$$

A ray incident with an angle larger than ϕ_c will be confined to the fibre and propagate in the fibre. A ray incident, at the core-cladding boundary, at the critical angle is called a **critical ray.** The critical ray makes an angle θ_c with axis of the fibre. It is obvious that rays with propagation angles larger than θ_c will not propagate in the fibre. Therefore, the angle θ_c is called the **critical propagation angle.** From the Δ^{le} ABC, it is seen that

$$\frac{AC}{AB} = \sin\phi_C, \qquad \text{Also, } \frac{AC}{AB} = \cos\theta_c$$

From the relation 4a $\sin\phi_c = n_2 / n_1$,

$$\cos\theta_c = n_2 / n_1 \qquad 5$$

$$\therefore \qquad \theta_c = \cos^{-1}(n_2/n_1) \qquad 6$$

Thus, only those rays which are refracted into the cable at angles $\theta_r < \theta_c$ will propagate in the optical fibre.

4.2 ACCEPTANCE ANGLE

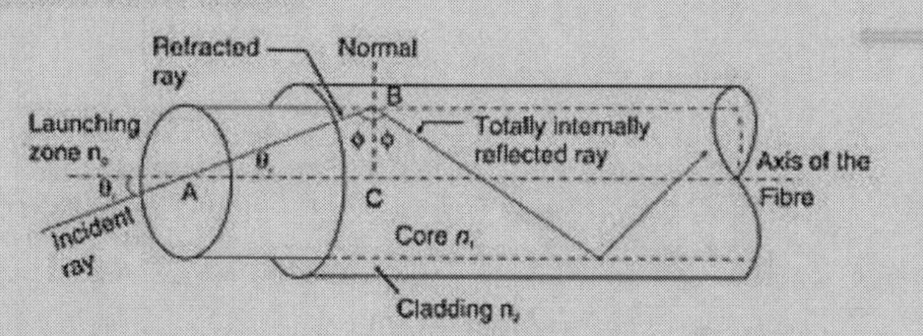

Fig 8 : Geometry for the calculation of acceptance angle of the fibre.

Let us again consider a step index optical fibre into which light is launched at one end, as shown in Fig. 24.8. Let the refractive index of the core be n_1 and the refractive index of the cladding be $n_2 (n_2 < n_1)$. Let n_o be the refractive index of the medium from which light is launched into the fibre.

Assume that a light ray enters the fibre at an angle θ_i to the axis of the fibre. The ray refracts at an angle θ_r and strikes the core-cladding interface at an angle ϕ. If ϕ is greater than critical angle ϕ_c, the ray undergoes total internal reflection at the interface, since $n_1 > n_2$. As long as the angle ϕ is greater than ϕ_c, the light will stay within the fibre.

Applying Snell's law to the launching face of the fibre, we get

$$\frac{\sin\theta_i}{\sin\theta_r} = \frac{n_1}{n_0} \qquad 7$$

If θ_i is increased beyond a limit, ϕ will drop below the critical value ϕ_c and the ray escapes from the sidewalls of the fibre. The largest value of θ_i occurs when $\phi = \phi_c$.

From the Δ^{le}ABC, it is seen that

$$\sin\theta_r = \sin(90^\circ - \phi) = \cos\phi \qquad 8$$

Using equation 8 into equation 7 we obtain

$$\sin\theta_i = \frac{n_1}{n_o}\cos\phi$$

When $$\phi = \phi_c, \quad \sin[\theta_{i_{max}}] = \frac{n_1}{n_o}\cos\phi_c \qquad 9$$

But $$\sin\phi_C = \frac{n_2}{n_1}$$

$$\therefore \qquad \cos\phi_C = \frac{\sqrt{n_1^2 - n_2^2}}{n_1} \qquad 10$$

Substituting the expression 10 into 9 we get

$$\sin\left[\theta_i(\max)\right] = \frac{\sqrt{n_1^2 - n_2^2}}{n_0} \qquad 11$$

Quite often the incident ray is launched from air medium, for which $n_o=1$.
Designating $\theta_i(max)= \theta_o$, equation 11 may be simplified to

$$\sin\theta_0 = \sqrt{n_1^2 - n_2^2}$$

$$\therefore \qquad \theta_o = \sin^{-1}\left[\sqrt{n_1^2 - n_2^2}\right] \qquad 12$$

The angle θ_o is called the **acceptance angle** of the fibre. *Acceptance angle is the maximum angle that a light ray can have relative to the axis of the fibre and propagate down the fibre.* Thus, only those rays that are incident on the face of the fibre making angles less than θ_o will undergo repeated total internal reflections and reach the other end of the fibre. Obviously, larger acceptance angles make it easier to launch light into the fibre.

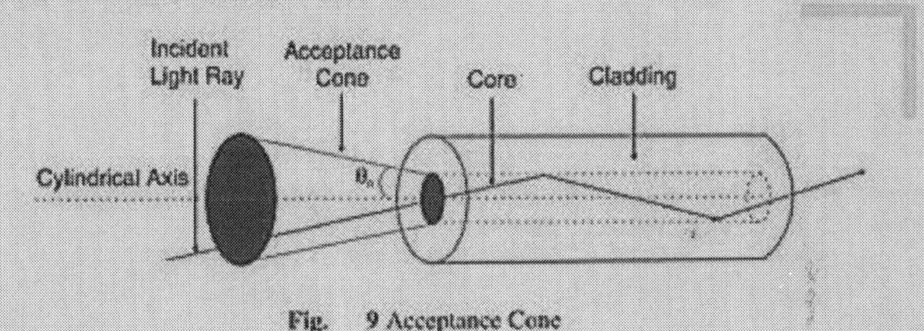

Fig. 9 Acceptance Cone

In three dimensions, the light rays contained within the cone having a full angle $2\theta_o$ are accepted and transmitted along the fibre (see Fig. 9). Therefore, the cone is called the **acceptance cone.**

Light incident at an angle beyond θ_o refracts through the cladding and the corresponding optical energy is lost.

5 FRACTIONAL REFRACTIVE INDEX CHANGE

The fractional difference Δ between the refractive indices of the core and the cladding is known as *fractional refractive index change*. It is expressed as

$$\Delta = \frac{n_1 - n_2}{n_1} \qquad 13$$

This parameter is always positive because n_1 must be larger than n_2 for the total internal reflection condition. In order to guide light rays effectively through a fibre, $\Delta \ll 1$. Typically, Δ is of the order of 0.01.

6 NUMERICAL APERTURE

The main function of an optical fibre is to accept and transmit as much light from the source as possible. The light gathering ability of a fibre depends on the numerical aperture. The acceptance angle and the fractional refractive index change determine the numerical aperture of fibre.

The numerical aperture (NA) is defined as the sine of the acceptance angle.

$$\sin\theta_0 = \sqrt{n_1^2 - n_2^2}$$

$$\therefore \quad NA = \sqrt{n_1^2 - n_2^2} \qquad 14$$

$$n_1^2 - n_2^2 = (n_1 + n_2)(n_1 - n_2) = \left(\frac{n_1 + n_2}{2}\right)\left(\frac{n_1 - n_2}{n_1}\right)2n_1$$

Approximating $\frac{n_1 + n_2}{2} \approx n_1$, we can express the above relation as $\left(n_1^2 - n_2^2\right) = 2n_1^2\Delta$. It gives

$$NA = \sqrt{2n_1^2\Delta}$$

$$\therefore \quad NA = n_1\sqrt{2\Delta} \qquad 15$$

Numerical aperture determines the light gathering ability of the fibre. It is a measure of the amount of light that can be accepted by a fibre. It is seen from equ. 14 that NA is dependent only on the refractive indices of the core and cladding materials and does not depend on the physical dimensions of the fibre. The value of NA ranges from 0.13 to 0.50. A large NA implies that a fibre will accept large amount of light from the source

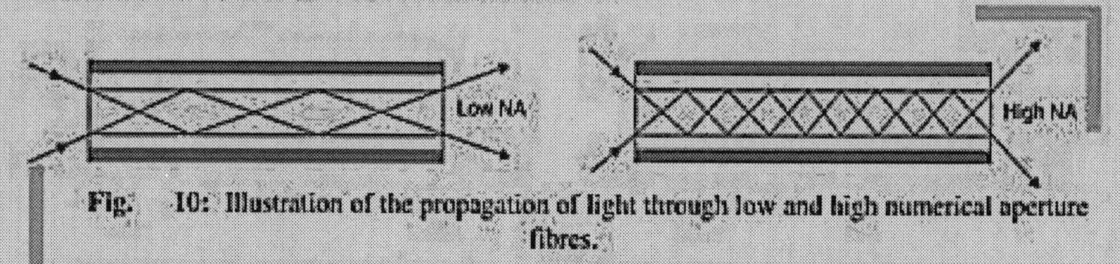

Fig. 10: Illustration of the propagation of light through low and high numerical aperture fibres.

7 SKIP DISTANCE AND NUMBER OF TOTAL INTERNAL REFLECTIONS

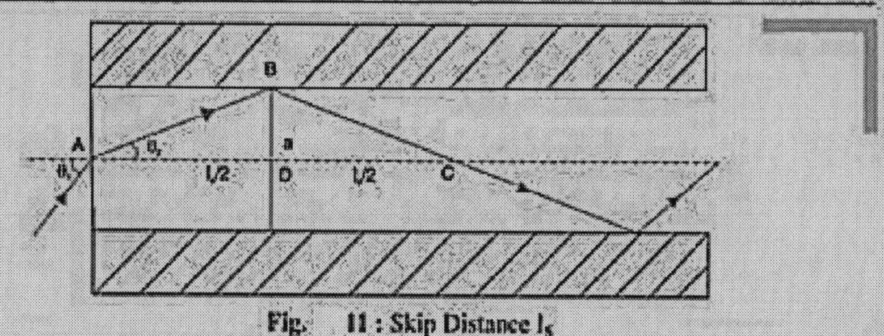

Fig. 11 : Skip Distance l_s

We shall now calculate the number of total internal reflections that a light ray undergoes as it travels through an optical fibre of length L. Let a be the radius of the fibre. Let a light ray be incident on one end of the fibre at an angle θ_1 to the axis and refract into the fibre at an angle θ_2, as shown in Fig. 11. The ray undergoes the first reflection at B. The distance $AC = l_s$ is known as the **skip distance** and represents the distance between two successive reflections of the ray. In the Δ^{le}ABD,

$$AD = \tfrac{1}{2}AC = \frac{l_s}{2}, \; BD = a \text{ and } \angle BAD = \theta_2.$$

$$\tan\theta_2 = \frac{BD}{AD} = \frac{a}{l_s/2}$$

or

$$l_s = \frac{2a}{\tan\theta_2} \qquad 16$$

In terms of the incidence angle θ_1, the above equation may be rewritten as

$$l_s = 2a\left[\frac{\cos\theta_2}{\sin\theta_2}\right] \quad \text{or} \quad l_s^2 = (2a)^2\left[\frac{1}{\sin^2\theta_2} - 1\right]$$

Using equ.(24.7) into the above expression, we obtain

$$l_s^2 = (2a)^2\left[\left(\frac{n_1}{n_a\sin\theta_1}\right)^2 - 1\right]$$

or

$$l_s = 2a\left[\left(\frac{n_1}{n_a\sin\theta_1}\right)^2 - 1\right]^{1/2} \qquad 17$$

In case of air, $n_a = 1$ and

$$l_s = 2a\left[\left(\frac{n_1}{\sin\theta_1}\right)^2 - 1\right]^{1/2} \qquad 17a$$

The number of total internal reflections in the total fibre length is given by

$$N = \frac{\textit{Total length of the cable}}{\textit{The distance travelled during one reflection}} = \frac{L}{l_s}$$

∴

$$N = \frac{L\tan\theta_2}{2a} \qquad 18a$$

Also,

$$N = \frac{L}{2a\left[\left(\frac{n_1}{\sin\theta_1}\right)^2 - 1\right]^{1/2}} \qquad 18b$$

For example, if $n_1 = 1.50$, $\theta_1 = 30^o$ and $a = 25\ \mu m$, equ.(24.18b) gives $l_s = 141\ \mu m$. Alternately, computing the value of θ_2 and using it in equ.(24.18a), we obtain $l_s = 141\ \mu m$. Therefore, if light travels through a length of 1 m of the optical fibre of the above specifications, it is reflected 7092 times.

8 MODES OF PROPAGATION

Light propagates as an electromagnetic wave through an optical fibre and its propagation is governed by Maxwell's equations. Complete understanding of propagation of light waves through optical fibres requires a thorough understanding of solution of these equations in the context of optical fibres. When a plane electromagnetic wave propagates in free space, it travels as a transverse electromagnetic wave. The electric field and magnetic field components associated with the wave are perpendicular to each other and also perpendicular to the direction of propagation. It is known as a TEM wave. When the light ray is guided through an optical fibre, it propagates in different types of modes. Each of these guided modes consists of a variety of electromagnetic field configurations, such as transverse electric (TE), transverse magnetic (T M) and hybrid modes. Hybrid modes are combination of transverse electric and magnetic modes.

In simple terms these *modes can be visualised as the possible number of allowed paths of light* in an optical fibre (see Fig.24.6). The paths are all zigzag paths excepting the axial direction. Though the rays having propagation angles between $\theta = 0^{o}$ and $\theta = \theta_c$ will be in a position to undergo total internal reflections, all of them will not however propagate along the optical fibre. Only a certain ray directions are allowed. As a zigzag ray gets repeatedly reflected at the walls of the fibre, phase shift occurs. Consequently, the waves travelling along certain zigzag paths will be in phase and undergo constructive interference, while the waves coursing along certain other paths will be out of phase and diminish due to destructive interference. The light ray paths along which the waves are in phase inside the fibre are known as **modes**. Each mode is a pattern of electric and magnetic field distributions that is repeated along the fibre at equal intervals. The number of modes propagating in a fibre increases as θ_c or Δ increases. Increasing the core refractive index increases the number of propagating modes. On the other hand, increasing the clad refractive index decreases the number of propagating modes. The number of modes that a fibre will support depends on the ratio d / λ, where d is the diameter of the core and λ is the wavelength of the wave being transmitted. The zero order ray travels along the axis is known as the *axial ray*.

Note that each mode carries a portion of the light from the input signal.

Types of modes:

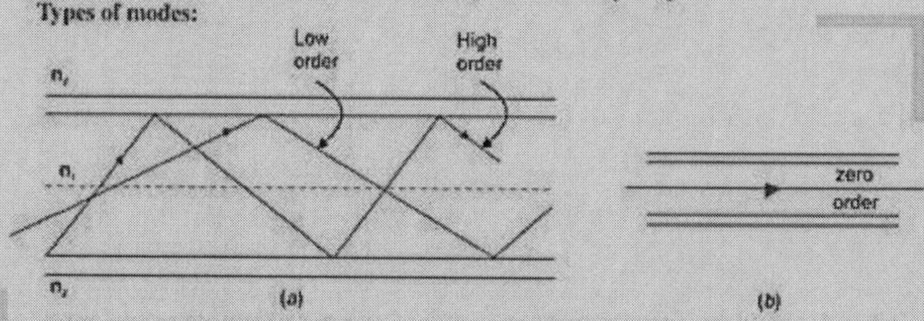

Fig. 12 : (a) Low and High-order ray paths in a multimode fibre. (b) Axial ray in a single mode fibre

In a fibre of fixed thickness, the modes that propagate at angles close to the critical angle ϕ_C (i.e., critical propagation angle θ_c) are **higher order modes**, and modes that propagate with angles larger than the critical angle (i.e., lower than the critical propagation angle) are **lower order modes** (see Fig. 12). In case of lower order modes, the fields are concentrated near the center of the fibre. In case of higher order modes, the fields are distributed more towards the edge of the wave-guide and tend to send light energy into the cladding. This energy is lost ultimately. The higher order modes have to traverse longer paths and hence take larger time than the lower order modes to cover a given length of the fibre. Thus, the higher order modes arrive at the output end of the fiber later than the lower order modes.

9 TYPES OF RAYS

The rays that propagate through an optical fibre can be classified into two categories: (*i*) meridional rays and (*ii*) skew rays.

1. Meridional ray: A ray that propagates through the fibre undergoing total internal reflection is called meridional ray. It passes through the longitudinal axis of the fibre core 24.13a The propagation of meridional rays is possible only in the TM or TE modes.

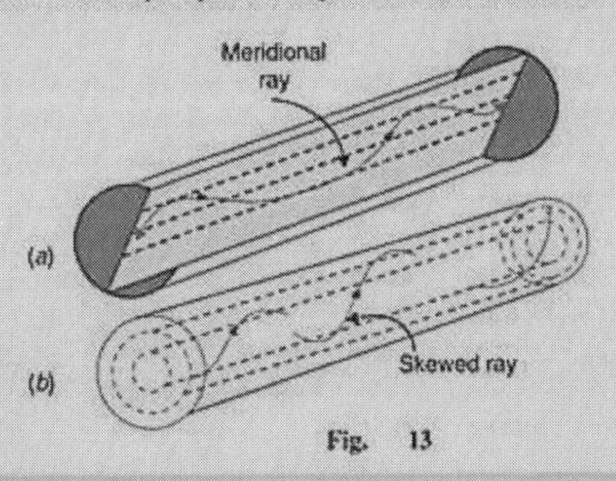

Fig. 13

2. Skew ray: The ray that describes angular helical path along the fibre is called a skew ray (see Fig. 13b). These rays do not pass through the axis of the core. These rays are propagated in either hybrid EH or HE modes. Some of these modes produce losses through leakage of radiation. In real situations, the skew rays constitute a substantial portion of the total number of guided rays. They tend to propagate only in the annular region near the outer surface of the core and do not fully utilize the core as the medium. However, they are complementary to the meridional rays and increase the light gathering capacity of the fibre.

10 CLASSIFICATION OF OPTICAL FIBRES

Optical fibres are differently classified into various types basing on different parameters.

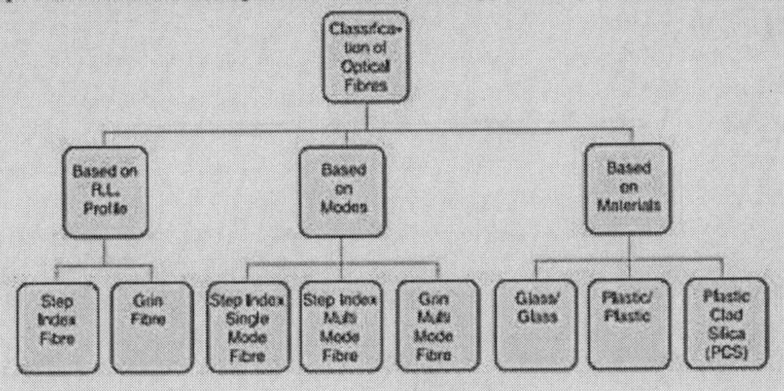

A. Classification basing on refractive index profile:

Refractive index profile of an optical fibre is a plot of refractive index drawn on one of the axes and the distance from the core axis drawn on the other axis (see Fig. 14). Optical fibres are classified into the following two categories on the basis of refractive index profile.

1. Step index fibres and **2. Graded index (GRIN) fibres**.

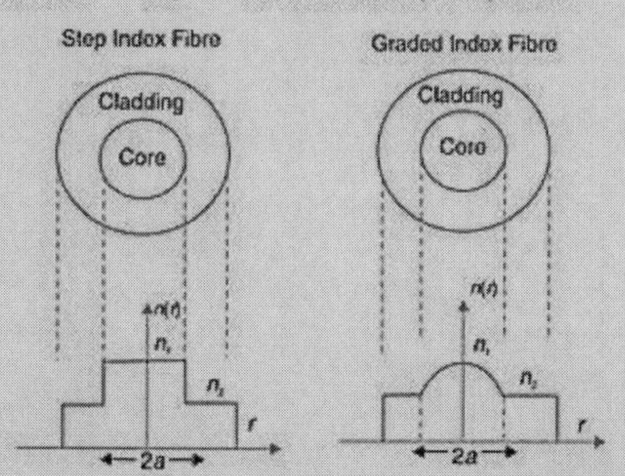

Fig. 14 : Classification of optical fibres based on R.I. profile (*a*) Step index fibre (*b*) GRIN fibre

Step index refers to the fact that the refractive index of the core is constant along the radial direction and abruptly falls to a lower value at the cladding and core boundary (see Fig. 14a). In case of GRIN fibres, the refractive index of the core is not constant but varies smoothly over the diameter of the core (see Fig. 14b). It has a maximum value at the center and decreases gradually towards the outer edge of the core. At the core-cladding interface the refractive index of the core matches with the refractive index of the cladding. The refractive index of the cladding is constant.

B. Classification basing on the modes of light propagation:

On the basis of the modes of light propagation, optical fibres are classified into two categories as

1. Single mode fibres (SMF) and **2. Multimode fibres (MMF)**.

A **single mode fibre** (SMF) has a smaller core diameter and can support only one mode of propagation. On the other hand, a **multimode fibre** (MMF) has a larger core diameter and supports a number of modes.

Thus, on the whole, the optical fibres are classified into three types:

- Single mode step-index (SMF) fibre
- Multimode step-index (MMF) fibre
- Graded index (multimode) (GRIN) fibre.

C. Classification basing on materials:

On the basis of materials used for core and cladding, optical fibres are classified into three categories.

1. Glass/glass fibres (glass core with glass cladding)
2. Plastic/plastic fibres (plastic core with plastic cladding)
3. PCS fibres (polymer clad silica)

We now study the detailed structure and characteristics of the three types of optical fibres.

11.1 SINGLE MODE STEP INDEX FIBRE

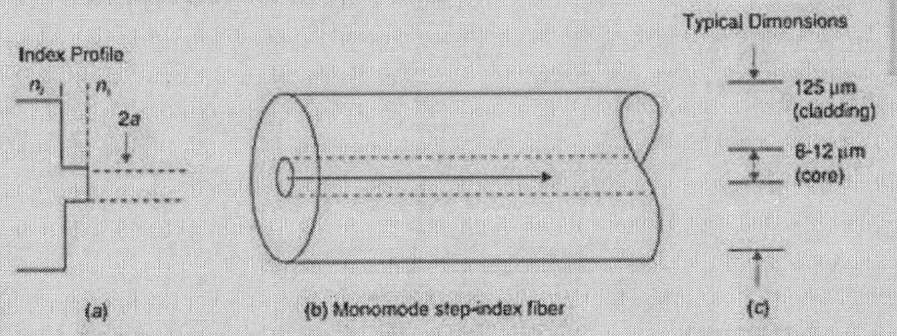

Fig. 15: Single mode step index fibre (a) R.I. profile (b) ray paths (c) typical dimensions

Structure

A single mode step index fibre has a very fine thin core of diameter of 8 μm to 12 μm (see Fig. 15 c). It is usually made of germanium doped silicon. The core is surrounded by a thick cladding of lower refractive index. The cladding is composed of silica lightly doped with phosphorous oxide. The external diameter of the cladding is of the order of 125 μm. The fibre is surrounded by an opaque protective sheath. The refractive index of the fibre changes abruptly at the core-cladding boundary, as shown in Fig. 15 (a). The variation of the refractive index of a step index fibre as a function of radial distance can be mathematically represented as

$$n(r) = n_1 \, [r < a \quad \text{inside core}] \\ = n_2 \, [r > a \quad \text{in cladding}] \qquad 19$$

Propagation of light in SMF

Light travels in SMF along a single path that is along the axis (Fig. 15b). Obviously, it is the zero order mode that is supported by a SMF. Both Δ and NA are very small for single mode fibres. This relatively small value is obtained by reducing the fibre radius and by making Δ, the relative refractive index change, to be small. The low NA means a low acceptance angle. Therefore, light coupling into the fibre becomes difficult. Costly laser diodes are needed to launch light into the SMF.

11.2 MULTIMODE STEP INDEX FIBRE

Structure

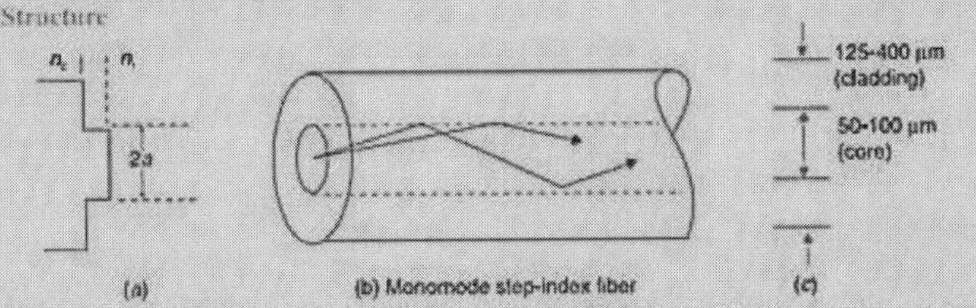

Fig. 16: Multimode step index fibre (a) R.I. Profile (b) Ray paths (c) typical dimensions.

A multimode step index fibre is very much similar to the single mode step index fibre except that its core is of larger diameter. The core diameter is of the order of 50 to 100 μm, which is very large compared to the wavelength of light. The external diameter of cladding is about 150 to 250 μm (Fig. 16 c).

Propagation of light in MMF

Multimode step index fibres allow finite number of guided modes. The direction of polarization, alignment of electric and magnetic fields will be different in rays of different modes. In other words, many zigzag paths of propagation are permitted in a MMF. The path length along the axis of the fibre is shorter while the other zigzag paths are longer. Because of this difference, the lower order modes reach the end of the fibre earlier while the high order modes reach after some time delay (Fig. 16b).

11.3 GRADED INDEX (GRIN) FIBRE

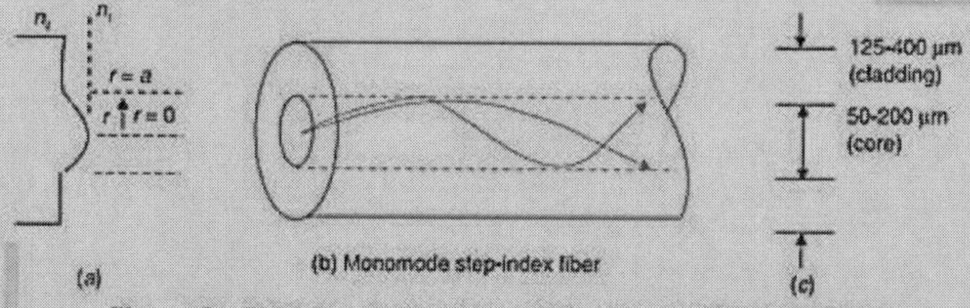

Fig. 17 : GRIN fibre (*a*) R.I. Profile (*b*) Ray paths (*c*) typical dimensions

A graded index fibre is a multimode fibre with a core consisting of concentric layers of different refractive indices. Therefore, the refractive index of the core varies with distance from the fibre axis. It has a high value at the centre and falls of with increasing radial distance from the axis. A typical structure and its index profile are shown in Fig. 17 (a). Such a profile causes a periodic focussing of light propagating through the fibre. The size of the graded index fibre is about the same as the step index fibre. The variation of the refractive index of the core with radius measured from the center is given by

$$n(r)=\begin{cases} n_1\sqrt{1-\left[2\Delta\left(\frac{r}{a}\right)^{\alpha}\right]}, & r<a \text{ inside core} \\ n_2, & r>a \text{ in cladding} \end{cases} \quad 20$$

where n_1 is maximum refractive index at the core axis, a the core radius, and α the grading profile index number which varies from 1 to ∞. When $\alpha = 2$, the index profile is parabolic and is preferred for different applications.

Propagation of light

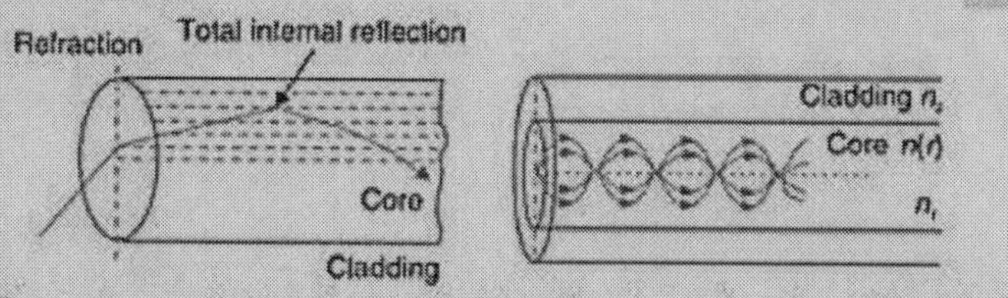

Fig. 18 : (a) An expanded ray diagram showing refraction at the various high to low index interfaces within graded index fibre, giving an overall curved ray path. (b) Light transmission in a graded index fibre.

As a light ray goes from a region of higher refractive index to a region of refractive index, it is bent away from the normal. The process continues till the condition for total internal reflection is met. Then the ray travels back towards the core axis, again being continuously refracted (Fig. 18a). The turning around may take place even before reaching the core-cladding interface. Thus, continuous refraction is followed by total internal reflection and again continuous refraction towards the axis. In the graded index fibre, rays making larger angles with the axis traverse longer path but they travel in a region of lower refractive index and hence at a higher speed of propagation. Consequently, all rays traveling through the fibre, irrespective of their modes of travel, will have almost the same optical path length and reach the output end of the fibre at the same time (see Fig. 18b).

In case of GRIN fibres, the acceptance angle and numerical aperture decrease with radial distance from the axis. The numerical aperture of a graded index fibre is given by

$$NA = \sqrt{n^2(r) - n_2^2} = n_1 (2\Delta)^{\frac{1}{2}} \sqrt{1 - \left(\frac{r}{a}\right)^2}$$

$$= n_1 \sqrt{2\Delta\left[1 - \left(\frac{r}{a}\right)^2\right]} \qquad (21)$$

12 MATERIALS

Optical fibres are fabricated from glass or plastic which are transparent to optical frequencies. Step index fibres are produced in three common forms – (*i*) a glass core cladded with a glass having a slightly lower refractive index, (*ii*) a silica glass core cladded with plastic and (*iii*) a plastic core cladded with another plastic. Generally, the refractive index step is the smallest for all glass fibres, a little larger for the plastic clad silica (PCS) fibres and the largest for all plastic construction.

12.1 ALL GLASS FIBRES

The basic material for fabrication of optical fibres is silica (SiO_2). It has a refractive index of 1.458 at $\lambda = 850\,nm$. Materials having slightly different refractive index are obtained by doping the basic silica material with small quantities of various oxides. If the basic silica material is doped with germania (GeO_2) or phosphorous pentoxide (P_2O_5), the refractive index of the material increases. Such materials are used as core materials and pure silica is used as cladding material in these cases. When pure silica is doped with boria (B_2O_3) or fluorine, its refractive index decreases. These materials are used for cladding when pure silica is used as core material. Examples of fiber compositions are

- SiO_2 core – $B_2O_3.SiO_2$ cladding
- $GeO_2.SiO_2$ core – SiO_2 cladding

The glass optical fibres exhibit very low losses and are used in long distance communications.

12.2 ALL PLASTIC FIBRES

In these fibres, perspex (PMMA) and polysterene are used for core. Their refractive indices are 1.49 and 1.59 respectively. A fluorocarbon polymer or a silicone resin is used as a cladding material. A high refractive index difference is achieved between the core and the cladding materials. Therefore, plastic fibres have large NA of the order of 0.6 and large acceptance angles up to 77°. The main advantages of the plastic fibres are low cost and higher mechanical flexibility. The mechanical flexibility allows the plastic fibres to have large cores, of diameters ranging from 110 to 1400 μm. However, they are temperature sensitive and exhibit very high loss. Therefore, they are used in low cost applications and at ordinary temperatures (below 80°C). Examples of plastic fiber compositions are

- Polysterene core — $n_1 = 1.60$ — NA = 0.60
  - Methyl methacrylate cladding — $n_2 = 1.49$
- Polymethyl methacrylate core — $n_1 = 1.49$ — NA = 0.50
  - cladding made of its copolymer — $n_2 = 1.40$

12.3 PCS FIBRES

The plastic clad silica (PCS) fibres are composed of silica cores surrounded by a low refractive index transparent polymer as cladding. The core is made from high purity quartz. The cladding is made of a silicone resin having a refractive index of 1.405 or of perfluoronated ethylene propylene (Teflon) having a refractive index of 1.338. Plastic claddings are used for step-index fibres only. The PCS fibres are less expensive but have high losses. Therefore, they are mainly used in short distance applications.

13 V-NUMBER

Let us consider a narrow beam of monochromatic light launched on the front end of a step-index fibre, at an angle less than the acceptance angle of the fibre. Let the wavelength of the light be λ_o and the diameter of the fiber be d. It appears to us from the ray concept that all the rays contained in the beam propagate along the fibre, such that there can be infinite modes of propagation. However, in practice, only a limited number of modes of propagation are possible in an optical fibre. To understand the reason for this behaviour, we have to recall that phase changes occur as the light waves travel forward. The phase shift takes place due to two reasons – (*i*) due to optical path length traversed and (*ii*) due to total internal reflection at the core-cladding interface.

(*i*) When a wave travels a distance l in a medium of refractive index n_1, it undergoes a phase change δ_1 given by

$$\delta_1 = k\, n_1 l = \frac{2\pi\, l\, n_1}{\lambda} \qquad 22$$

where k is the propagation constant.

(*ii*) Whenever, a wave with component normal to the reflecting surface undergoes total internal reflection, the phase shift, δ_2, is given by

$$\delta_2 = 2\tan^{-1}\frac{\sqrt{n_1^2\cos^2\phi - n_2^2}}{n_1\sin\phi} \qquad 23$$

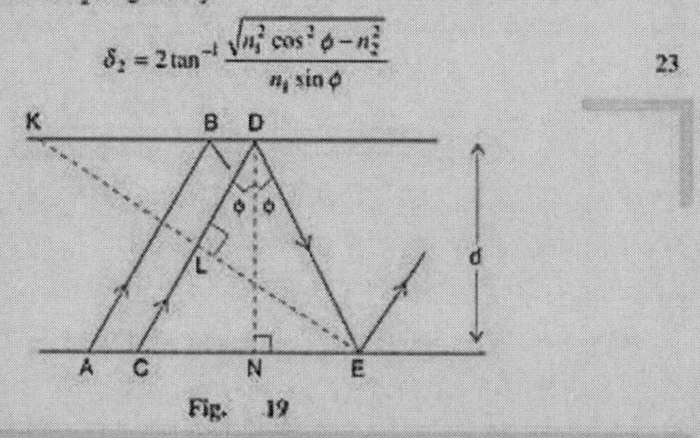

Fig. 19

In Fig. 19, AB and CD are parallel rays in an incident beam. The line KL is perpendicular to the propagation path of the rays AB, CD and hence represents a plane wavefront. The points K and L lying on the same wavefront will be in phase with each other. As the point E, which is on the reflected ray DE, lies on the wavefront KL, the points L and E must be in phase with each other. However, moving from the point L to E along the ray, we find that there occurs a phase shift given by

$$\delta = (LD + DE)\frac{2\pi\, n_1}{\lambda_o} - 2\delta_2 \qquad 24$$

The factor 2 in the above equation takes into account the two total internal reflections at D and E. In the $\Delta^{le}\, DNE$

$$\frac{DN}{DE} = \cos\phi \text{. Therefore, } DE = \frac{DN}{\cos\phi} = \frac{d}{\cos\phi}$$

Further, in the $\Delta^{le}\, LDE$

$$\frac{LD}{DE} = \cos 2\phi \text{. Therefore, } LD = DE\cos 2\phi$$

$$LD + DE = DE\,(1 + \cos 2\phi) = 2\,DE\cos^2\phi.$$

Therefore, $$LD + DE = 2\frac{d}{\cos\phi}\cos^2\phi = 2d\cos\phi$$

Using the above expression into equ.(24.23), we obtain

$$\delta = \frac{4d\pi\, n_1 \cos\phi}{\lambda_o} - 2\delta_2$$

Now the condition for the wave associated with the ray CD to propagate along the optical fibre is that the phase of the twice reflected wave must be the same as that of the incident wave. That is, the wave must interfere constructively with itself. If this phase condition is not satisfied, the wave would interfere destructively with itself and just die out. It means that the total phase shift must be equal to an integer multiple of 2π radians. Thus

$$\frac{4d\pi\, n_1 \cos\phi}{\lambda_o} - 2\delta_2 = 2\pi\, m$$

or $$m = \frac{2d\, n_1 \cos\phi_m}{\lambda_o} - \frac{\delta_2}{\pi} \qquad 25$$

where m is an integer that determines the allowed ray angles for propagation of the wave and ϕ_m is the value of ϕ corresponding to a particular value of m. In order to sustain total internal reflection,

$$\sin\phi_m \geq \frac{n_2}{n_1}$$

$$\therefore \quad \cos\phi_m \leq \frac{\sqrt{n_1^2 - n_2^2}}{n_1}$$

$$\therefore \quad m \leq \frac{2d\sqrt{n_1^2 - n_2^2}}{\lambda_o} - \frac{\delta_2}{\pi} \qquad 26$$

or $$m \leq \frac{2V}{\pi} - \frac{\delta_2}{\pi} \qquad 27$$

where V is given by

$$V = \frac{\pi\, d}{\lambda_o}\sqrt{n_1^2 - n_2^2} \qquad 28$$

V-number is more generally called normalized **frequency** of the fibre. Each mode has a definite value of V-number below which the mode is cut off. Equ. 28 can be written as

$$V = \frac{\pi d}{\lambda_o}(NA) \qquad 29$$

or

$$V = \frac{\pi d}{\lambda_o} n_1 \sqrt{2\Delta} \qquad 30$$

The maximum number of modes N_m supported by an SI fibre is given by

$$N_m = \frac{1}{2}V^2 \qquad 31$$

Thus, for $V = 10$, N_m is 50. When the normalized frequency V is less than 2.405, the fibre can support only one mode, which propagates along the axial length of the fibre, and the fibre becomes a single mode fibre. It means that for single mode transmission in a MMF, V must be less than 2.405. The wavelength at which the fibre becomes single mode is called **cutoff wavelength,** λ_c of the fibre. Using equ. 29 we can write

$$\lambda_c = \frac{\pi d}{2.405}(NA) \qquad 32$$

It is seen from the above equation that single mode property can be realized in a multimode fibre by decreasing the core diameter and/or decreasing Δ such that V < 2.405.

In case of GRIN fibres, for larger values of V,

$$N_m \cong \frac{V^2}{4} \qquad 33$$

14 FABRICATION

A number of techniques are available to produce all glass fibres. In one of the methods, known as the double crucible method, fibres are directly produced from the melt.

Double Crucible Technique

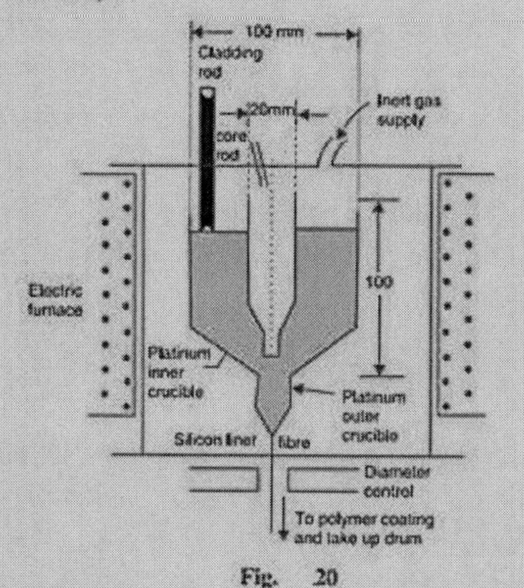

Fig. 20

The double crucible consists of two concentric platinum crucibles having thin orifices at the bottom. Raw material for the core-glass is placed in the inner crucible and the raw material for cladding is fed to the outer crucible. The double crucible arrangement is mounted vertically (Fig.24.20) in a furnace. The furnace is maintained at a suitable temperature to take the raw material into molten state. The fibres are drawn through the thin orifices at the bottom of the crucibles. As both the materials are drawn simultaneously, a filament of core glass surrounded by a tube of cladding glass is obtained in the process. The thickness of the fibre is monitored and the fibre is then coated with a polymer. Subsequently, it is passed through a plastic extrusion die to form a plastic sheath over the fibre.

15 LOSSES IN OPTICAL FIBRE

As a light signal propagates through a fibre, it suffers loss of amplitude and change in shape. The loss of amplitude is referred to as *attenuation* and the change in shape as *distortion*.

15.1 ATTENUATION

When an optical signal propagates through a fibre, its power decreases exponentially with distance. The loss of optical power as light travels down a fiber is known as **attenuation**. The attenuation of optical signal is defined as *the ratio of the optical output power from a fibre of length L to the input optical power.* If P_i is the optical power launched at the input end of the fibre, then the power P_o at a distance L down the fibre is given by

$$P_o = P_i\, e^{-\alpha L} \qquad 34$$

where α is called the **fibre attenuation coefficient** expressed in units of km^{-1}. Taking logarithms on both the sides of the above equation, we obtain

$$\alpha = \frac{1}{L} ln \frac{P_i}{P_o} \qquad (24.35)$$

In units of dB / km, α is defined through the equation

$$\therefore \qquad \alpha_{dB/km} = \frac{10}{L} \log \frac{P_i}{P_o} \qquad 36$$

In case of an ideal fibre, $P_o = P_i$ and the attenuation would be zero.

15.2 DIFFERENT MECHANISMS OF ATTENUATION

There are several loss mechanisms responsible for attenuation in optical fibres. They are broadly divided into two categories: *intrinsic* and *extrinsic* attenuation. Intrinsic attenuation is caused by substances inherently present in the fiber, whereas extrinsic attenuation is caused by external forces such as bending.

A. Intrinsic Attenuation

Intrinsic attenuation results from materials inherent to the fiber. It is caused by impurities present in the glass. During manufacturing, there is no way to eliminate all impurities. When a light signal hits an impurity in the fiber, either it is scattered or it is absorbed. Intrinsic attenuation can be further characterized by two components:

- **Material absorption**
- **Rayleigh scattering**

Absorption by material

Material absorption occurs as a result of the imperfection and impurities in the fiber and accounts for 3-5% of fiber attenuation. The most common impurity is the hydroxyl (OH-) molecule, which remains as a residue despite stringent manufacturing techniques. These radicals result from the presence of water remnants that enter the fiber-optic cable material through either a chemical reaction in the manufacturing process or as humidity in the environment. The natural impurities in the glass absorb light signal, and convert it into vibrational energy or some other form of energy. Hydroxyl radical ions(OH), and transition metals such as copper, nickel, chromium, vanadium and manganese have electronic absorption in and near visible part of the spectrum. Their presence causes heavy losses.

Even a highly pure glass absorbs light in specific wavelength regions. Strong electronic absorption occurs at UV wavelengths, while vibrational absorption occurs at IR wavelengths.

Losses due to impurities can be reduced by better manufacturing processes. In improved fibres, metal ions are practically negligible. The largest loss is caused by OH ions. These cannot be sufficiently reduced. The absorption of light either through intrinsic or impurity process constitutes a transmission loss because that much energy is subtracted from the light propagating through the fibre. The absorption losses are found to be at minimum at around 1.3 µm.

Unlike scattering, absorption can be limited by controlling the amount of impurities during the manufacturing process.

Rayleigh Scattering

Rayleigh scattering accounts for the majority (about 96%) of attenuation in optical fiber. The local microscopic density variations in glass cause local variations in refractive index. These variations, which are inherent in the manufacturing process and cannot be eliminated, act as obstructions and scatter light in all directions (Fig.24.21). This is known as *Rayleigh scattering*. The Rayleigh scattering loss greatly depends on the wavelength. It varies as $1/\lambda^4$ and becomes important at lower wavelengths. Thus, Rayleigh scattering sets a lower limit, on the wavelengths that can be transmitted by a glass fibre at 0.8 µm, below which the scattering loss is very high.

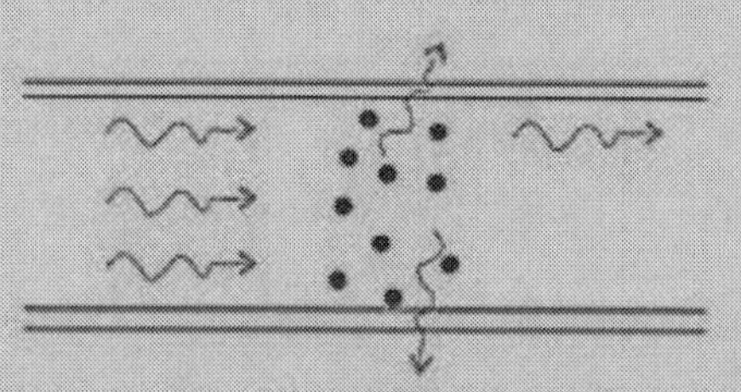

Fig. 21: Rayleigh scattering, showing attenuation of an incident stream of photons due to localized variations in refractive index.

Any wavelength that is below 800 nm is unusable for optical communication because attenuation due to Rayleigh scattering is high. At the same time, propagation above 1700 nm is not possible due to high losses resulting from infrared absorption.

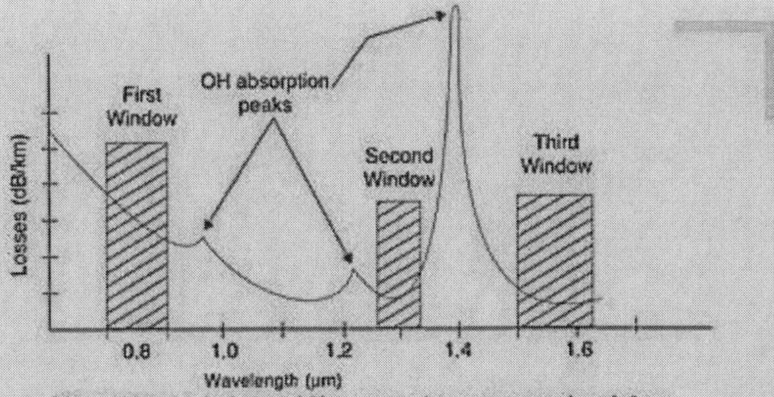

Fig. 22: A typical plot of fibre attenuation versus wavelength for a silica based optical fibre.

Fig. 22 shows the variation of attenuation with wavelength measured for a typical fiber-optic cable.

For better performance, the choice of wavelength **must** be based on minimizing loss and minimizing dispersion. Such windows are selected for communication purposes. It is seen from the attenuation curve that it has a minimum at around a particular band of optical wavelengths. The band of wavelengths at which the attenuation is a minimum is called **optical window** or **transmission window** or **low-loss window**. There are three **principal windows**. These correspond to wavelength regions in which attenuation is low and matched to the capability of a transmitter to generate light efficiently and a receiver to carry out detection.

λ (nm)	**Approx. loss (dB/km)**
820 - 880	2.2
1200 - 1320	0.6
1550 - 1610	0.2

From the above data it is seen that the range 1550 to 1610 is most preferable. From the point of view of dispersion, the low intramodal dispersion wavelength of about 1300nm is most suitable.

B. Extrinsic Attenuation or Bending losses

Extrinsic attenuation is caused by two external mechanisms: **macrobending or microbending**. Both of them cause a reduction of optical power. If a bend is imposed on an optical fiber, strain is placed on the fiber along the region that is bent. The bending strain affects the refractive index and the critical angle of the light ray in that specific area. As a result, the condition for total internal reflection is no longer satisfied. Hence, light traveling in the core can refract out, and loss occurs.

Macrobend losses

A **macrobend** is a large-scale bend that is visible. When a fibre is bent through a large angle, strain is placed on the fiber along the region that is bent. The bending strain will affect the refractive index and the critical angle of the light ray in that specific area. As a result, light traveling in the core can refract out, and loss occurs. (Fig. 24.23). To prevent macrobends, optical fiber has a *minimum bend radius* specification that should not be exceeded. This is a restriction on how much bend a fiber can withstand before experiencing problems in optical performance or mechanical reliability.

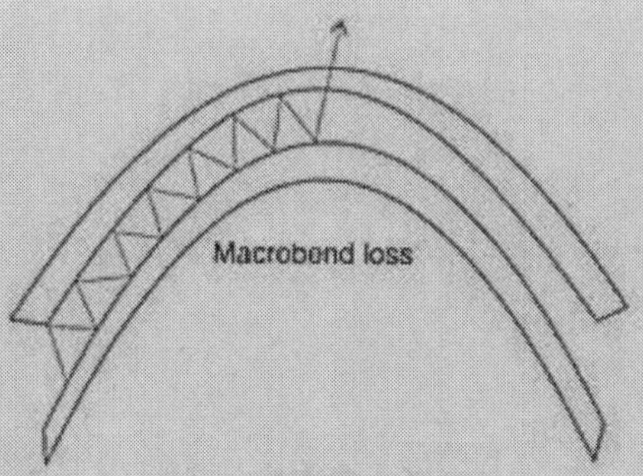

Fig. 23 : Macrobend loss

Microbend losses

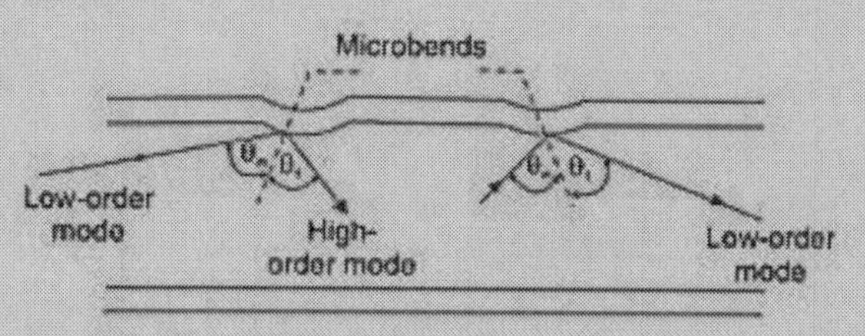

Fig. 24: Microbend losses

Microbend is a small-scale distortion. It is localized and generally indicative of pressure on the fiber. Microbending might be related to temperature, tensile stress, or crushing force. Microbending is caused by imperfections in the cylindrical geometry of fiber during the manufacturing process or installation processes. The bend may not be clearly visible upon inspection. Structural variations in the fibre, or fibre deformation, cause radiation of light away from the fibre (Fig. 24). Microbending may occur, for example, due to winding of optical fibre cable over spools. Light rays get scattered at the small bends and escape into the cladding. Such losses are known as microbend losses.

16 DISTORTION

In an optical fibre communication system, the information (signal) is coded in the form of discrete pulses of light, which are transmitted through the fibre. The light pulses are of a given width, amplitude and interval. The number of pulses that can be sent per unit time will determine the information capacity of the fibre. More information can be sent by optical cable when distinct pulses can be transmitted in more rapid succession. The pulses travel through the transmitting medium (i.e., optical fibre) and reach the detector at the receiving end. For the information to be retrieved at the detector, it is necessary that the optical pulses are well resolved in time. However, the light pulses broaden and spread into a wider time interval because of the different times taken by different rays propagating through the fibre. This phenomenon is known as **distortion** or **pulse dispersion**. Hence, even though two pulses may be well resolved at the input end, they may overlap on each other at the output end, as shown in Fig.24.25. It is obvious that the pulse broadening depends on the length of the travel of the pulses through the fibre. Hence, dispersion is expressed in units of **ns/km** (time/distance).

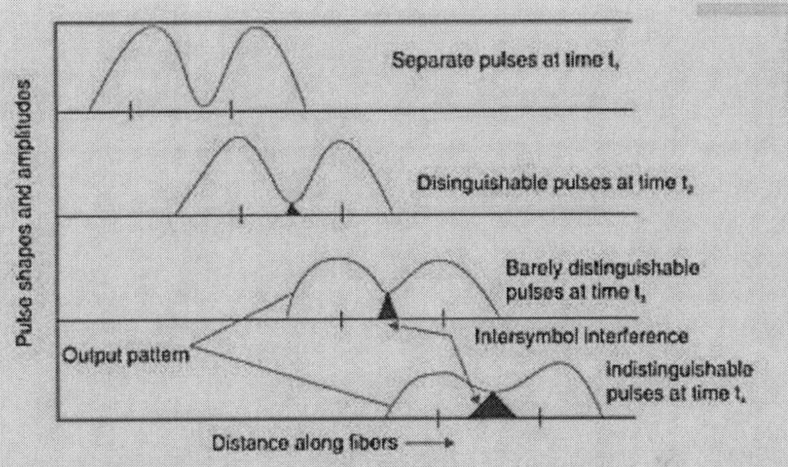

Fig. 25 : Distortion of the pulses traveling along a fibre

The following three different dispersion mechanisms determine the distortion of the signal in an optical fibre. They are

- Intermodal dispersion and
- Intramodal dispersion.

Intramodal dispersion is again divided into the following two types.

- Material dispersion
- Waveguide dispersion

16.1 INTERMODAL DISPERSION

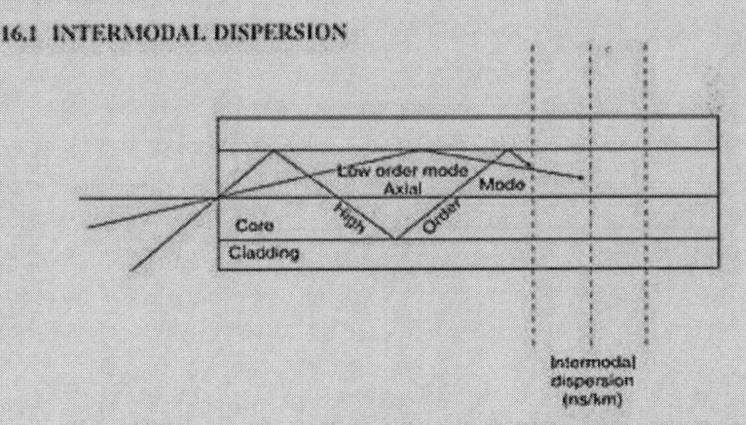

Fig. 26: Lower order modes reach the end of the fibre earlier while the high order modes reach after some time delay

Intermodal dispersion occurs as a result of the differences in the group velocities of the modes. For example, let us consider the propagation of a pulse through a multimode fibre. The power associated with the single pulse gets distributed into the various modes or paths guided by the fibre.

The lower order modes (rays reflected at larger angles) travel a greater distance than the higher order modes (lower angle rays). The path length along the axis of the fibre is shorter while the other zigzag paths are longer. Because of this difference, the lower order modes reach the end of the fibre earlier while the high order modes reach after some time delay. As a result, light pulses broaden as they travel down the fibre, causing signal distortion. The output pulses no longer resemble the input pulses (see Fig. 26). This type of distortion is known as **intermodal** or simply **modal dispersion**. This imposes limitation on the separation between successive pulses and thereby reduces the transmission rate and capacity.

Expression for total time delay due to modal dispersion in Step-Index fibre:

The total time delay between the arrival of the axial ray and the slowest ray, the one traveling the longest distance is

$$\Delta t = t_{max} - t_{min} \qquad 37$$

Referring to the Fig. 11, the time taken by a refracted ray to traverse the distance ABC of the fibre would be

$$t' = \frac{AB + BC}{\upsilon} = \frac{n_1 AC}{c \cos\theta_r}$$

where $\upsilon = c/n_1$ is the speed of the light in the core. Since the ray path will repeat itself, the time taken by a ray to traverse a length L of the fibre is

$$t = \frac{n_1 L}{c \cos\theta_r} \qquad 38$$

The above relation shows that the time taken by a ray in the fibre core is a function of the angle θ_r. For the axial ray $\theta_r = 0$ and hence

$$t_{min} = \frac{n_1 L}{c} \qquad 39$$

In case of the ray that travels the longest path, $\theta_r = \theta_C$. Therefore,

$$t_{max} = \frac{n_1 L}{c \cos\theta_C}$$

Using equ.(24.6) into the above expression, we get

$$t_{max} = \frac{n_1^2 L}{n_2 c} \qquad 40$$

Therefore, making use of equations (24.40) and (24.39) into (24.37), we obtain

$$\Delta t = \frac{n_1 L}{c}\left[\frac{n_1}{n_2} - 1\right] \qquad 41$$

Using the equ.(24.13) for fractional refractive index change into the above equ.(24.41), we get

$$\Delta t = \frac{n_1 L}{c}\left[\frac{\Delta}{1-\Delta}\right] \qquad 42$$

We can also express the relation (24.41) in the following form.

$$\Delta t = \frac{n_1 L}{c}\left[\frac{n_1 - n_2}{n_2}\right] = \frac{n_1 L}{c}\left[\frac{n_1 - n_2}{n_2}\right]\left[\frac{n_1 + n_2}{n_1 + n_2}\right]$$

$$= \frac{n_1 L}{c}\left[\frac{(n_1^2 - n_2^2)}{n_2(n_1 + n_2)}\right] = \frac{n_1 L}{c}\frac{(n_1^2 - n_2^2)}{2n_1 n_2}$$

or
$$\Delta t = \frac{L}{2n_2 c}(NA)^2 \qquad 43$$

It is seen from the equ. 43 that the time delay is proportional to the square of the value of NA. Therefore, a large NA fibre allows more modes of propagation of light, which will result in greater modal dispersion. A smaller NA limits the number of modes, hence reduces dispersion. It is further seen that the intermodal dispersion does not depend upon the spectral width of the source. It follows that a light pulse from an ideal monochromatic source would still get broadened.

16.2 INTRAMODAL DISPERSION

Intramodal dispersion is the spreading of light pulse within a single mode. The two main causes of intramodal dispersion are (*a*) material dispersion and (*b*) waveguide dispersion.

a. Material Dispersion: Glass is a dispersive medium. A light pulse is a wave packet, composed of a group of components of different wavelengths. The different wavelength components will propagate at different speeds along the fibre (Fig. 27). The short wavelength components travel slower than long wavelength components, eventually causing the light pulse to broaden. This type of distortion is known as **material dispersion.** It is often called the **chromatic dispersion.** Obviously, the spectral width of the source determines the extent of material dispersion.

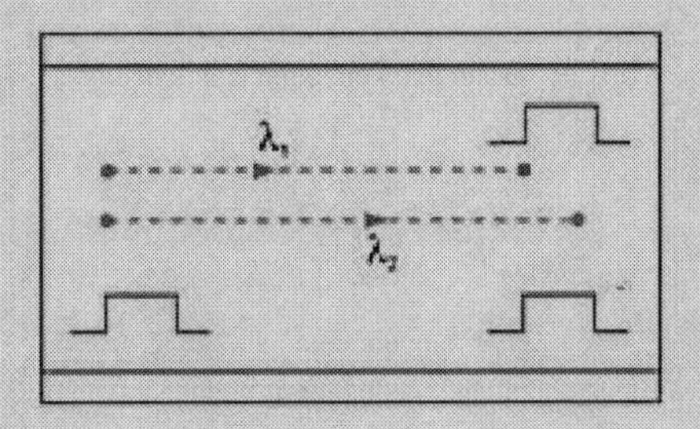

Fig. 27 : The different wavelength components propagate at different speeds along the fibre

Expression for time delay due to material dispersion:

Let us consider a plane wave propagating in fiber core. It is represented by $\psi \propto \exp(kx - \omega t)$. The wave number k is given by

$$k = \frac{2\pi}{\lambda} = \frac{2\pi}{\lambda_o}\cdot\frac{\lambda_o}{\lambda} = \frac{2\pi}{\lambda_o}\cdot n = \frac{2\pi\nu}{c}\cdot n$$

or
$$k = \frac{\omega n}{c} \qquad 44$$

and
$$\omega = 2\pi\nu = \frac{2\pi}{\lambda}c \qquad 45$$

A wave packet of finite spread of wavelengths travels with group velocity v_g is given by

$$v_g = \frac{d\omega}{dk}$$

$$\therefore \quad \frac{1}{v_g} = \frac{dk}{d\omega} = \frac{d}{d\omega}\left(\frac{\omega n}{c}\right) = \frac{n}{c} + \frac{\omega}{c}\frac{dn}{d\omega} = \frac{1}{c}\left[n + \omega\frac{dn}{d\omega}\right]$$

But

$$\frac{dn}{d\omega} = \frac{dn}{d\lambda}\cdot\frac{d\lambda}{d\omega} = -\frac{\lambda^2}{2\pi c}\cdot\frac{dn}{d\lambda}$$

$$\therefore \quad \frac{1}{v_g} = \frac{1}{c}\left[n - \frac{\omega\lambda^2}{2\pi c}\frac{dn}{d\lambda}\right] = \frac{1}{c}\left[n - \lambda\frac{dn}{d\lambda}\right]$$

As the signal propagates through the fibre, each spectral component can be assumed to travel independently and to undergo a time delay per unit length in the direction of propagation, which is given by

$$t_{mat} = \frac{L}{v_g} = \frac{L}{c}\left[n - \lambda\frac{dn}{d\lambda}\right] \qquad 46$$

The pulse spread Δt_{mat} for a source of spectral width $\Delta\lambda$ is found by differentiating the equ.(24.46) with respect to λ and then multiplying by $\Delta\lambda$. Thus,

$$\Delta t_{mat} = \frac{dt_{mat}}{d\lambda}\Delta\lambda = -\frac{L\lambda}{c}\frac{d^2n}{d\lambda^2}\Delta\lambda = D_{mat}(\lambda)L\lambda \qquad 47$$

where $D_{mat}(\lambda)$ is the mateial dispersion.

$$D_{mat}(\lambda) = -\frac{\lambda}{c}\frac{d^2n}{d\lambda^2} \qquad 48$$

From the equ.(24.48) it is seen that the material dispersion can be reduced either by choosing sources with narrower spectral range or by operating at longer wavelengths. To cite an example, an LED operating at 820 nm and having a spectral width of 38 nm results in dispersion of about 3 ns/km in a certain fibre. In the same fibre, dispersion can be reduced to 0.3 ns/km using a laser diode operating at 1140 nm and having a spectral width of 3 nm. Thus, using a more and more monochromatic source operating at higher wavelength, the material dispersion is reduced.

b. Wave-guide Dispersion: Waveguide dispersion arises from the guiding properties of the fibre. The group velocities of modes depend on the wavelength. Hence, the effective refractive index for any mode varies with wavelength. It is equivalent to the angle between the ray and the fibre axis varying with wavelength which subsequently leads to a variation in the transmission times for the rays and hence dispersion (see Fig.24.28). Waveguide dispersion is generally small in MMF, but it is important in SMF.

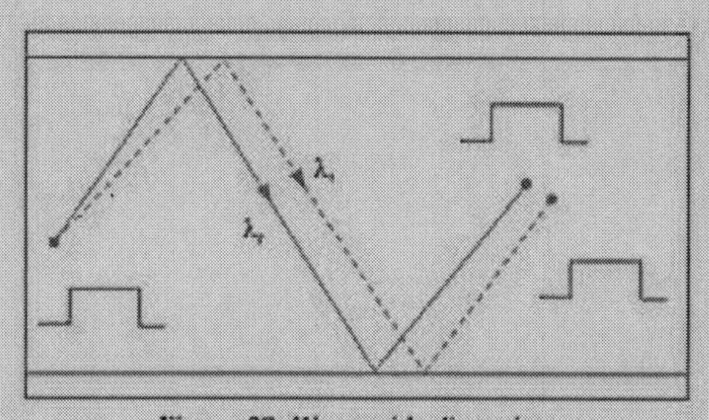

Fig. 28: Wave guide dispersion

The intermodal distortion can be reduced if graded index fibre is used. In case of a graded index fibre, the refractive index is larger at the center and it gradually decreases away from the center. A pulse traveling along the axis of the fibre, travels along a shorter path but it takes longer time to reach the end of the fibre since it is traveling through a medium of higher refractive index. On the other hand, the pulse traveling away from the axis travels a longer distance but takes lesser time since it is traveling through a medium of lower refractive index. As a result both the pulses reach the end of the fibre simultaneously. Thus, using a GRIN fibre can reduce the problem of intermodal dispersion. Low NA fibres exhibit smaller dispersion. Dispersion may be restricted by a careful selection of low NA fibre and a narrow spectral width fibre.

In a MMF, all three pulse spreading mechanism exist simultaneously. In case of SMF, only material and wave-guide dispersion exist.

16.3 TOTAL DISPERSION

All the above three dispersions contribute pulse spreading during signal transmission through an optical fibre. The total dispersion introduced by an optical fibre is given by the root mean square value of all the three dispersions. Thus,

$$(\Delta t)_T = \sqrt{(\Delta t)^2_{intermodal} + (\Delta t)^2_{mat} + (\Delta t)^2_{wg}} \qquad 49$$

17 BANDWIDTH

It is learnt in the above section that various dispersion mechanisms cause broadening of the information signal in time domain. If the pulses spread more, they can interfere with the adjacent pulses resulting in *Inter Symbol Interference* or ISI in short and there can be so much of ISI that it becomes impossible to distinguish between the individual pulses. Therefore, for a given broadening, the pulses have to be separated by a minimum time interval in order to avoid overlapping of the pulses. This would determine the ultimate information-carrying capacity of the system. When the pulse separation is increased, the data transfer rate decreases. Thus, broadening of pulses puts an upper limit on the rate of pulse transmission. To a first approximation, it may be taken that the bandwidth in hertz is equal to the digital bit rate. Thus, $B_T = \frac{1}{\tau} = B$, where τ is the input pulse duration. In other words, the maximum allowable transmission rate is called **bandwidth**. In practice, the fibre bandwidth is expressed in terms of MHz.km, a product of frequency and distance. This is known as **bandwidth-distance product**, which specifies the usable bandwidth over a definite distance. With the increase in distance, different dispersion effects would increase in the optical fibre and as a result the usable bandwidth reduces. The attenuation per kilometer and the bandwidth-kilometer product are the important performance parameters of optical fibres.

18 CHARACTERISTICS OF THE FIBRES

A. Step-index single-mode fibre

- It has a very small core diameter, typically of about 10 μm.
- Its numerical aperture is very small.
- It supports only one mode in which the entire light energy is concentrated.
- A single mode step index fibre is designed to have a V number between 0 and 2.4.
- Because of a single mode of propagation, loss due to intermodal dispersion does not exist. With careful choice of material, dimensions, and wavelength, the total dispersion can be made extremely small.

- The attenuation is least.
- The single mode fibres carry higher bandwidth than multimode fiber.
- It requires a monochromatic and coherent light source. Therefore, laser diodes are used along with single mode fibres.

Advantages

- No degradation of signal
- Low dispersion makes the fibre suitable for use with high data rates. Single-mode fiber gives higher transmission rate and up to 50 times more distance than multimode.
- Highly suited for communications.

Disadvantages

- Manufacturing and handling of SMF are more difficult.
- The fibre is costlier.
- Launching of light into fibre is difficult.
- Coupling is difficult.

Applications

- Used as under water cables

B. Step-index multi-mode fibre

- It has larger core diameter, typically ranging between 50-100 μm.
- The numerical aperture is larger and it is of the order of 0.3.
- Larger numerical aperture allows more number of modes, which causes larger dispersion. The dispersion is mostly intermodal.
- Attenuation is high.
- Incoherent sources like LEDs can be used as light sources with multimode fibres.

Advantages

- The multimode step index fibre is relatively easy to manufacture and is less expensive.
- LED or laser source can be used.
- Launching of light into fibre is easier.
- It is easier to couple multi-mode fibres with other fibres.

Disadvantages

- Has smaller bandwidth.
- Due to higher dispersion data rate is lower and transmission is less efficient.
- It is less suitable for long distance communications.

Applications

- Used in data links.

C. Graded-index multi-mode fibre

- Core diameter is in the range of 50-100 μm.
- Numerical aperture is smaller than that of step-index multimode fibre.
- The number of modes in a graded index fibre is about half that in a similar multimode step-index fibre.
- Has medium attenuation.
- Intermodal dispersion is zero, but material dispersion is present.
- Has better bandwidth than multimode step-index fibre.

Advantages

- Either an LED or a laser can be used as the source of light with GRIN fibres.

Mechanical fiber splices are designed to be quicker and easier to install, but there is still the need for stripping, careful cleaning and precision cleaving. The fiber ends are aligned and held together by a precision-made sleeve, often using a clear index-matching gel that enhances the transmission of light across the joint. Such joints typically have higher optical loss and are less robust than fusion splices, especially if the gel is used. All splicing techniques involve the use of an enclosure into which the splice is placed for protection afterward.

(*i*) **V- groove splice technique**

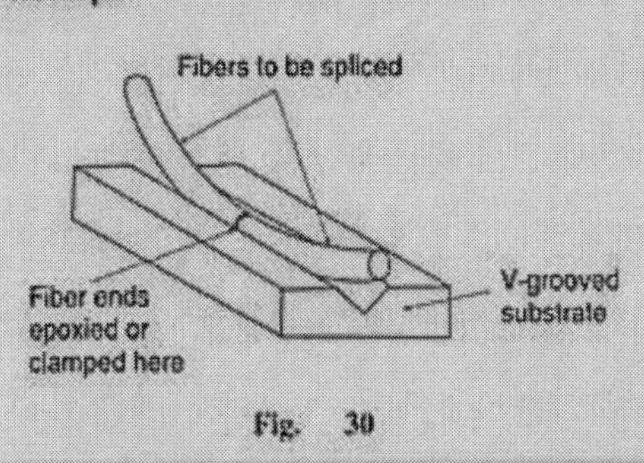

Fig. 30

The V-block is the simplest mechanical splice. The bared fibres to be joined are placed in the groove (Fig. 24.30). Angular alignment is particularly well controlled. The two fibres can slide in the groove until they touch. They are then epoxied permanently into position, so end-separation errors are minimal. If the epoxy is index matched to the fibre, even small gaps can be tolerated with little loss. Lateral misalignment would be negligible in the groove if both fibres had the same core and cladding diameters. A cover plate can be placed over the V-block to protect the splice further.

(*ii*) **Elastomer splice technique**

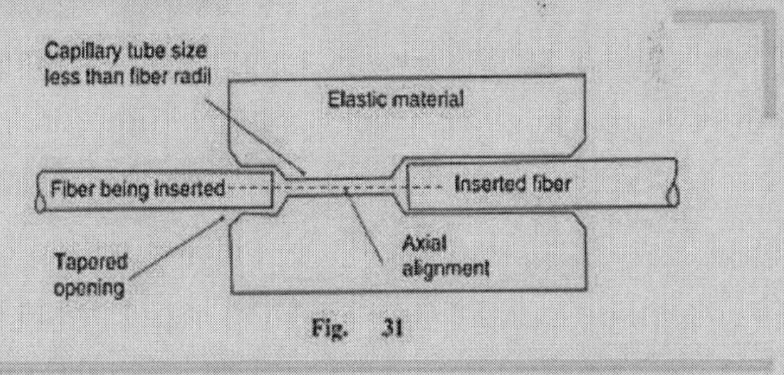

Fig. 31

Another splice is essentially a precision sleeve made with elastomeric materials. The elastomer is an elastic material usually made into a cylinder with an opening along its axis. The groove is a little smaller than the fibre but accepts and centers it by expanding slightly when the fibre is inserted (Fig. 31). The fibres are inserted from both the ends of the cylinder and touch near its midpoint. The slice can be epoxied for permanent connection. An external splice holder is used for full protection of the splice.

20 APPLICATIONS

Transmission of light via an optical fibre has a wide variety of applications. We discuss here some of the applications. Broadly, optical fibres have three different applications, apart from other miscellaneous applications.

a. They are used for illumination and short distance transmission of images.

b. They are used as wave-guides in telecommunications.

c. They are used in fabricating a new family of sensors.

20.1 ILLUMINATION AND IMAGE TRANSMISSION

A large number of fibres whose ends are bound together, ground and polished, form flexible bundles. One of the ends of the bundle acts as an input end while the other acts as an output end. If the relative positions of the fibre terminations at both the ends are not the same, and if no attempt is made to align the fibres in an orderly array, the bundle is said to be an *incoherent bundle*. In such a case, there would not be any correlation in the positions of the fibre terminations at one end of the bundle with that at the other end of the bundle. The primary function of such bundles is simply to conduct light from one region to another. Such **flexible light carriers** are relatively easy to make and inexpensive. They are used for illumination purpose.

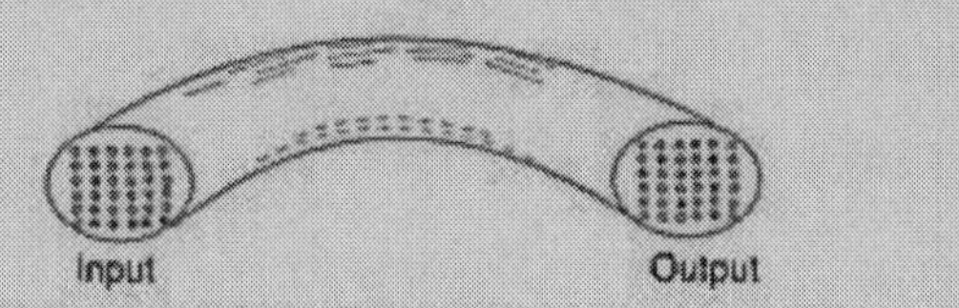

(a) Incoherent bundle-image of letter L scrambled as dark spots at the out put

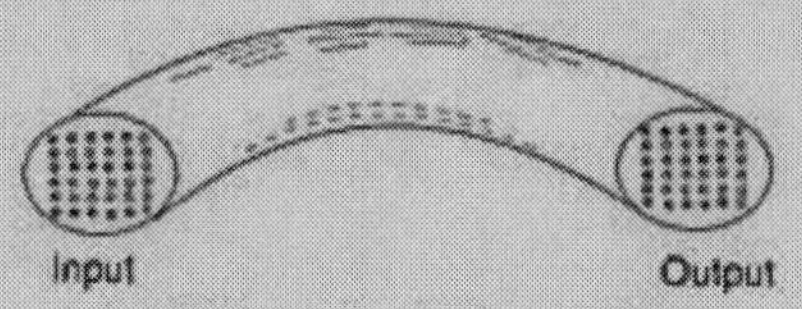

(a) Coherent bundle-image of letter E

Fig. 32 : Fibre Optic bundles

When the fibres are carefully arranged so that their terminations occupy the same relative positions in both of the bound ends of the bundle, the bundle is said to be *coherent*. Such a bundle is capable of transmitting undistorted images to a distant place. When one end of such a **flexible image carrier** is placed face down flat on an illuminated surface, a point-by-point image of the surface appears at the other end (Fig. 32).

When the fibres are carefully arranged so that their terminations occupy the same relative positions in both of the bound ends of the bundle, the bundle is said to be *coherent*. Such a bundle is capable of transmitting undistorted images to a distant place. When one end of such a **flexible image carrier** is placed face down flat on an illuminated surface, a point-by-point image of the surface appears at the other end.

Endoscopes

The most important application of the coherent bundles is in diagnostic field as an optical endoscope. An endoscope is an optical instrument which facilitates visual inspection of internal parts of a human body. It is also called a fiberoscope. It requires about 10,000 fibres forming a bundle of 1 mm diameter and it can resolve objects with a separation of 70 µm. By allowing direct viewing of what was formerly hidden, a fiberoscope has become a vital diagnostic tool for industry and medicine. The broncho-fiberoscope, gastrointestinal fiberoscope, laparoscope etc are the endoscopes used in medical diagnosis.

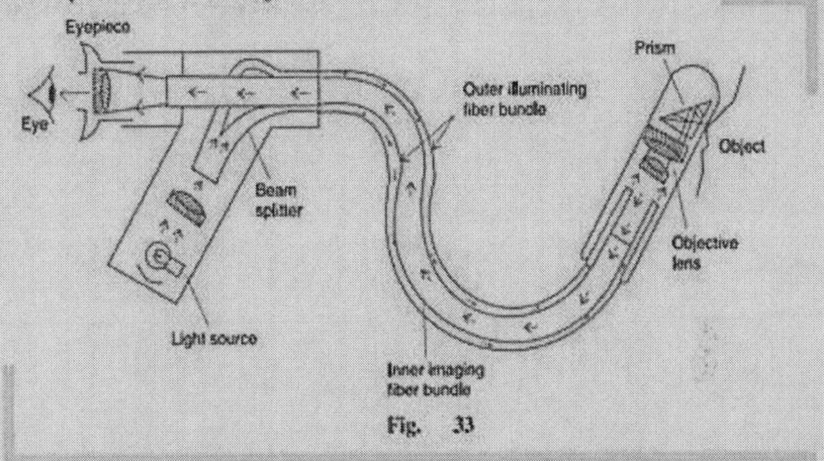

Fig. 33

Fig. 33 shows the schematic diagram of a flexible endoscope. The endoscopes are designed using low quality, large diameter and short silica fibres. There are two fiber bundles in an endoscope. One of them is used to illuminate the interior of the body and the other is used to collect the reflected light from the illuminated area. A telescope system is added in the internal part of endoscope for obtaining a wider field of view and better image quality. At the object end, there is an assembly of objective lens and prism which are kept in a transparent glass cover and at the viewing end, there is an eye lens. The input end of the endoscope contains a powerful light source. The light rays are focused and coupled to the illuminating fiber bundle. The light rays are finally incident on the surface of the object under study. The light rays reflected from the object surface are received by the objective lens through a prism and are transmitted through the imaging fibre bundle to the viewing end of the scope. Here the eye piece reconstructs the image of the object and one can view the image of the surface of the object. Endoscope pictures can be recorded on a videotape recorder.

20.2 OPTICAL COMMUNICATIONS

Traditionally, electronic communications were carried out by sending electrical signals through copper cables, coaxial cables or waveguides. In recent years optical fibres are being used, where light signals replace electrical signals. A basic communications system consists of a transmitter, a receiver and an information pathway. Normally, the information to be communicated is a non-electrical message, which is to be converted first into an electrical form. The conversion is done by a transducer. For example, a microphone converts sound waves into currents. Similarly, a video camera converts images into currents. These electrical messages are of low frequency and cannot be transmitted directly. Therefore, they are superposed on a carrier wave of very high frequency. The process of imposing a message signal on a carrier wave is called modulation. Two different techniques of modulation are available. In analog modulation a continuous wave carries the message. In digital modulation message is transmitted in discrete form using binary digits. The message travels along the transmission channel and is received at the receiver. The receiver demodulates the modulated

The block diagram Fig. 34 illustrates a typical communications system. The transmitter consists of a light source supported by necessary drive circuits. A transducer converts a non-electrical message into an electrical signal and is fed to a light source. The light source is a miniature source, either a light emitting diode or a semiconductor laser. In either case, light is emitted in the IR range with a wavelength of 850 nm (0.85 μm), 1300 nm (1.3 μm) or 1550 nm (1.55 μm). The light waves are modulated with the signal. By varying the intensity of the light beam from the laser diode or LED, analog modulation is achieved. By flashing the laser diode or LED on and off at an extremely fast rate, digital modulation is achieved. A pulse of light represents the number 1 and the absence of light at a specified time represents zero. A message can be transmitted by a particular sequence of these 1s or 0s. If the receiver is programmed to recognize such digital patterns, it can reconstruct the original message. Though the digital modulation requires more complicated equipment such as encoders and decoders and also more bandwidth than analog modulation, it allows greater transmission distance with the same power. This is a great advantage and hence digital modulation has become popular and widely used nowadays. The transmitter feeds the analog or digitally modulated light wave to the transmission channel, namely optical fibre link. The optical signal travelling through the fibre will get attenuated progressively and distorted due to dispersion effects. Therefore, repeaters are to be used at specific intervals to regenerate the signal. At the end of the fibre, an output coupler directs the light from the fibre onto a semiconductor photodiode, which converts the light signals to electrical signals. The photodetector converts the light waves into electrical signals which are then amplified and decoded to obtain the message. The output is fed to a suitable transducer to convert it into an audio or video form.

Applications

Optical fibre communications systems can be broadly classified into two groups: (*i*) local and intermediate range systems where the distances involved are small and (*ii*) long-haul systems where cables span large distances.

(*i*) Local area networks: The local area network (LAN) is a computer oriented communication system. LAN operates over short distances of about 1 to 2 km. It is multiuser oriented system. In LAN, a number of computer terminals are interconnected over a common channel allowing each computer to use data and programs from any other. An optical data bus offers a great reduction in cost and increases enormously the information handling capacity.

(*ii*) Long-haul communication: One of the most important applications of fibre optic communication is long-haul communication. Long-haul communication systems are used for long distances, 10 km or more. Telephone cables connecting various countries come under this category. A rather sophisticated long-haul network is the NSFNET which links six supercomputer centres throughout U.S.A.

22 MERITS OF OPTICAL FIBRES

Optical fibres have many advantageous features that are not found in conducting wires. Some of the important advantages are given here.

1. Cheaper: Optical fibres are made from silica (SiO_2) which is one of the most abundant materials on the earth. The overall cost of a fibre optic communication is lower than that of an equivalent cable communication system.

2. Smaller in size, lighter in weight, flexible yet strong: The cross section of an optical fibre is about a few hundred microns. Hence, the fibres are less bulky. Typically, a RG-19/U coaxial cable weighs about 1100 kg/km whereas a PCS fibre cable weighs 6 kg/km only. Optical fibres are quite flexible and strong.

3. Not hazardous: A wire communication link could accidentally short circuit high voltage lines and the sparking occurring thereby could ignite combustible gases in the area leading to a great damage. Such accidents cannot occur with fibre links since fibres are made of insulating materials.

5. Immune to EMI and RFI: In optical fibres, information is carried by photons. Photons are electrically neutral and cannot be disturbed by high voltage fields, lightening, etc. Therefore, fibres are immune to externally caused background noise generated through electromagnetic interference (EMI) and radiofrequency interference (RFI).

6. No cross talk: The light waves propagating along the optical fibre are completely trapped within the fibre and cannot leak out. Further, light cannot couple into the fibre from sides. In view of these features, possibility of cross talk is minimized when optical fibre is used. Therefore, transmission is more secure and private.

7. Wider bandwidth: Optical fibres have ability to carry large amounts of information. While a telephone cable composed of 900 pairs of wire can handle 10,000 calls, a 1mm optical fibre can transmit 50,000 calls.

8. Low loss per unit length: The transmission loss per unit length of an optical fibre is about 4 dB/km. Therefore, longer cable-runs between repeaters are feasible. If copper cables are used, the repeaters are to be spaced at intervals of about 2 km. In case of optical fibres, the interval can be as large as 100 km and above.

22.1 DISADVANTAGES

Installation and maintenance of optical fibres require a new set of skills. They require specialized and costly equipment like optical time domain reflectometers etc. All this means heavy investment.

23 FIBRE OPTIC SENSORS

Fibre optic sensors are transducers, which generally consist of a light source coupled with an optical fibre and a light detector held at the receiver-end. The fibres used could be either multimode or single mode type. The sensors can be used to measure pressure, temperature, strain, the acoustic field, magnetic field, etc physical parameters. The advantages of these sensors are that they are lighter, occupy lesser volume and are cheaper.

The optical fibre merely carries the light beam in some of the sensors and in others the fibre itself acts as the sensor. We study here a few typical examples of the sensors.

23.1 TEMPERATURE SENSORS

(a) Intensity modulated sensor

Principle: In this type of sensor, temperature is measured by the modulation of intensity of the reflected light from a target, a silicon layer. The operation of the temperature sensor is based on the 1 μm wavelength light-absorption characteristics of silicon as a function of temperature. Depending on the temperature, the amount of light absorbed by the silicon layer varies. The change in intensity of the reflected light is proportional to the change in temperature.

Construction: Fig. 35 illustrates a temperature sensor with a multimode fibre. The fibre is coated at one end with a thin silicon layer.

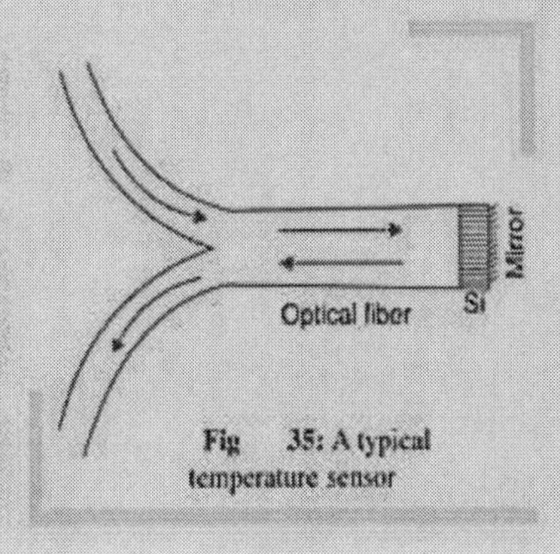

Fig 35: A typical temperature sensor

The silicon layer is in turn coated with a reflective coating at the back. The silicon layer acts as the sensing element.

Working: The light from a light source is launched into the fibre from one of the ends of one of its branches (see Fig. _ 35). It passes first through the fibre and then through the silicon layer. The mirror coating at the other end of the silicon layer reflects the light back which again travels through the silicon layer. The reflected light emerges out through another branch of multimode fibre and is collected by a photodetector. The amount of the reflected light is converted into voltage by the photodetector. The absorption of light by the silicon layer varies with temperature and the variation modulates the intensity of the light received at the detector. Temperature measurements can be made with a sensitivity of 0.001°C.

(*b*) Phase modulated sensor

Principle: This temperature sensor is based on phase variation resulting due to the variation of refractive index of the optical fibre under the influence of temperature.

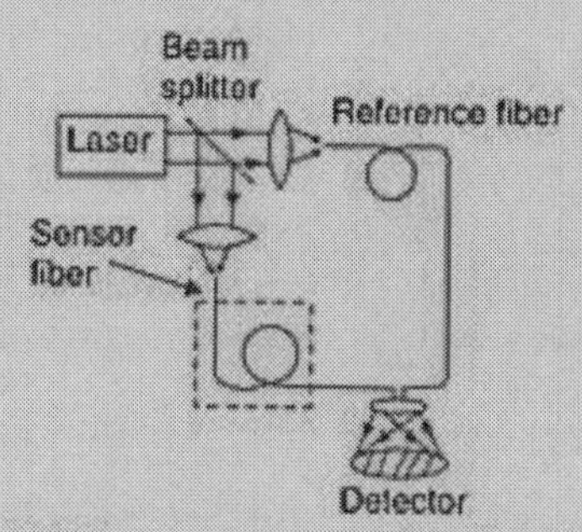

Fig. 36: Temperature sensor using phase variations

Construction: Fig. 36 shows a single mode fibre sensor arranged in what is known as the Mach-Zehnder arrangement. A light source produces light. A beam splitter divides the light into two parts and sends light through the sensing fibre and the reference fibre. Light passing out of the two fibre elements is fed to a detector, which measures the difference in phase of the two light waves. Accurate measurements of the temperature may be obtained from these patterns.

Working: The light from the source is divided into two parts by the beam splitter. One part is allowed through sensor fibre, and the other part is passed through the reference fibre. Light rays entering the fibres are coherent and have the same phase. Prior to heating, the optical path lengths of the two fibre elements are same and hence both the outputs will be in phase. When the sensor fibre is subjected to heating, the temperature causes a change in the refractive index of the optical fibre. Therefore, the light coming out of the two fibres at the other end will have phase difference due to difference in optical path difference caused by the heating. When the rays are superposed, they interfere and interference pattern will be observed. As temperature increases, the phase difference between the two outputs increases and is observed as a displacement of the fringe pattern. By determining the fringe displacement, we can determine the magnitude of temperature.

23.2 DISPLACEMENT SENSOR

Principle: The basic principle employed in displacement sensor consists of using an adjacent pair of fibre optic elements, one to carry light from a remote source to an object whose displacement or motion is to be measured and the other to receive the light reflected from the object and carry it back to a remote photodetector.

Construction: An optical fibre without jacket is placed held between two corrugated blocks, as shown in Fig.24.38. Light from a source is divided into two parts by a beam splitter. One part is allowed through the fibre that is held between the blocks, which acts as a sensor element, and the other part is passed through an exactly identical fibre, which acts as a reference element. Photodetectors measure the intensity of transmitted light. A comparator detects the difference between the light intensities.

Working: When a force is applied on the upper corrugated block, the fibre is pressed and microbend losses are introduced in the fibre. The microbendings produce mode coupling such that energy of one mode is transferred to other higher modes. Also, higher modes are converted into leaky modes which reduce the amount of energy transmitted though the fibre. The changes in the light intensity due to these losses are detected by a photodetector and compared with that of the light coming out of the reference element. The change in intensity is related to the force and hence is a measure of the applied force.

23.4 LIQUID LEVEL DETECTOR

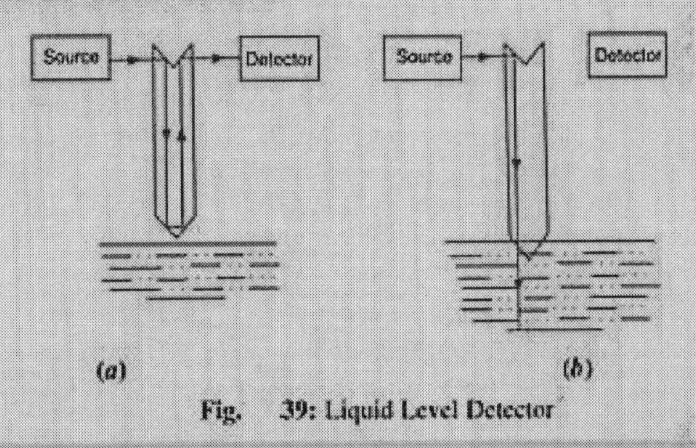

Fig. 39: Liquid Level Detector

Principle: The liquid level detector described here is based on the principle of total internal reflection.

Construction: A simple liquid level detector is shown in Fig. 39. A notch is made at one end of a multimode optical fibre and its other end is chamfered as shown in Fig. 39. A light source sends light on to the fibre and a photodetector on the other side registers light emerging out from the fibre.

Working: The optical fibre is arranged at the desired height in a vessel. The refractive index of the fibre is chosen to be less than that of the liquid whose level is to be detected. Light from the light source is made to be incident on one of the inclined faces of the notch. The light turns through 90° and travels through the fibre. On reaching the chamfered end of the fibre, it gets internally reflected at the fibre-air boundary, if the liquid is below the desired level. Then, it is again turned through 90° at the opposite face, travels back through the fibre to be turned once again through 90° and is detected at the detector (Fig. 39 a).

When the liquid rises and touches the fibre end, total internal reflection ceases and the light is transmitted into the liquid. Hence, the photodetector does not receive any light (Fig. 39 b). Thus, an indication of the liquid level is obtained at the detector.

WORKED-OUT EXAMPLES

Example 1: In an optical fibre, the core material has refractive index 1.43 and refractive index of clad material is 1.4. Find the propagation angle.

Solution: $$\cos\theta_C = \frac{n_2}{n_1} = \frac{1.40}{1.43} = 0.979$$

Therefore, propagation angle $\theta_C = \cos^{-1}(0.979) = 11.8^\circ$

Example 2: In an optical fibre, the core material has refractive index 1.6 and refractive index of clad material is 1.3. What is the value of critical angle? Also calculate the value of angle of acceptance cone.

Solution: Critical angle is given by $\sin\phi_c = \frac{n_2}{n_1} = \frac{1.3}{1.6} = 0.8125$

$$\therefore \phi_c = 54.3^\circ$$

Acceptance angle $$\theta_0 = \sin^{-1}\left[\sqrt{n_1^2 - n_2^2}\right] = \sin^{-1}\left[\sqrt{1.6^2 - 1.3^2}\right]$$

$$= \sin^{-1}(0.87)$$
$$= 60.5^\circ$$

Angle of acceptance cone $= 2\theta_0 = 121^\circ$

Example 3: Calculate the numerical aperture and acceptance angle of an optical fibre from the following data:

$$\mu_1(core) = 1.55 \text{ and } \mu_2(cladding) = 1.50$$

Solution: $$NA = \sqrt{n_1^2 - n_2^2} = \sqrt{1.55^2 - 1.50^2} = \sqrt{0.153} = 0.391.$$

Acceptance angle $$\theta_0 = \sin^{-1}\left[\sqrt{n_1^2 - n_2^2}\right] = \sin^{-1}\left[\sqrt{1.55^2 - 1.50^2}\right] = 23.02^\circ$$

Example 4: What is the numerical aperture of an optical fibre cable with a clad index of 1.378 and a core index of 1.546?

Solution: $$NA = \sqrt{n_1^2 - n_2^2} = \sqrt{1.546^2 - 1.378^2} = \sqrt{0.491} = 0.70$$

Example 5: A fibre cable has an acceptance angle of 30° and a core index of refraction of 1.4. Calculate the refractive index of the cladding.

Solution: $$\sin\theta_o = \sqrt{n_1^2 - n_2^2}$$

$$\therefore \sin^2\theta_o = n_1^2 - n_2^2$$

$$n_2^2 = n_1^2 - \sin^2\theta_o = (1.4)^2 - \sin^2 30^\circ = 1.96 - 0.25$$
$$= 1.71$$

$$\therefore \quad n_2 = 1.308$$

Example 6: Calculate the angle of acceptance of a given optical fibre, if the refractive indices of the core and the cladding are 1.563 and 1.498 respectively.

Solution:

$$\sin\theta_o = \sqrt{n_1^2 - n_2^2} = \sqrt{(1.563)^2 - (1.498)^2} = 0.4461$$

$$\theta_o = \sin^{-1}(0.4461) = 26.49^\circ$$

Example 7: Calculate the fractional index change for a given optical fibre if the refractive indices of the core and the cladding are 1.563 and 1.498 respectively.

Solution: Fractional index change $\Delta = \frac{n_1 - n_2}{n_1} = \frac{1.563 - 1.498}{1.563} = \frac{0.065}{1.563} = 0.0415$

Example 8: Calculate the refractive indices of the core and the cladding material of a fiber from the following data:

Numerical aperture (NA) = 0.22 and $\Delta = 0.012$

where Δ is the fractional refractive index change.

Solution:

$$NA = n_1\sqrt{2\Delta}$$

$$0.22 = n_1\sqrt{2\times 0.012} = 0.155\ n_1.$$

$$\therefore \quad n_1 = \frac{0.22}{0.155} = 1.42$$

$$\Delta = \frac{n_1 - n_2}{n_1} \therefore \frac{1.42 - n_2}{1.42} = 0.012 \therefore n_2 = 1.42 - 1.42\times 0.012 = 1.403$$

Example 9: Find the fractional refractive index and numerical aperture for an optical fibre with refractive indices of core and cladding as 1.5 and 1.49 respectively.

Solution:

$$\Delta = \frac{n_1 - n_2}{n_1} = \frac{1.5 - 1.49}{1.5} = 0.0067$$

$$NA = n_1\sqrt{2\Delta} = 1.5\sqrt{2\times 0.0067} = 0.174$$

Example 10: A step-index fibre is made with a core of refractive index 1.52, a diameter of 29 μm and a fractional difference index of 0.0007. It is operated at a wavelength of 1.3 μm. Find the V-number and the number of modes that the fibre will support.

Solution:

$$V = \frac{\pi d}{\lambda_o} n_1\sqrt{2\Delta} = \frac{3.143\times 29\times 10^{-6}\,m}{1.3\times 10^{-6}\,m} \times 1.52\sqrt{2\times 0.0007} = 4.049$$

$\therefore$ Number of modes, $N = \frac{1}{2}V^2 = \frac{1}{2}(4.049)^2 = 8$ **modes**

Example 11: A step-index fibre is with a core of refractive index 1.55 and cladding of refractive index 1.51. Compute the intermodal dispersion per kilometer of length of the fibre and the total dispersion in a 15 km length of the fibre.

Solution:

$$\Delta t = \frac{n_1 L}{c}\left[\frac{n_1}{n_2} - 1\right] = \frac{1.55\times 10^3\,m}{3\times 10^8\,m/s}\left[\frac{1.55}{1.51} - 1\right] = 138\ \text{ns/km}.$$

Total dispersion for 15 km length = $\Delta t \times 15$ km = (138 ns/km)×15 km = **2.07 μs.**

22. What are the different types of attenuation losses in an optical fibre? Discuss the absorption losses.
23. Describe various mechanisms of attenuation in optical fibres.
24. Draw the diagram for an optical fibre link and explain the function of each block.
25. List the main components of optical communication system. Describe the basic optical communication system.
26. Explain optical communication through block diagram. For long distance communication whether (*i*) mono-mode or multimode and (*ii*) step index or graded index fibre, which are preferable and why?
27. Discuss the advantages and disadvantages of optical fibres over conventional communication transmission media.
28. Explain with basic principle, the construction and working of any one type of optical fibre sensor.
29. Discuss any one application of an optical fibre as a sensor.

PROBLEMS FOR PRACTICE

1. An optical fibre has a core material of refractive index of 1.55 and cladding material of refractive index 1.50. The light is launched into the fibre from air. Calculate its numerical aperture.
2. The numerical aperture of an optical fibre is 0.39. If the difference in the refractive indices of the material of its core and the cladding is 0.05, calculate the refractive index of material of the core.
3. An optical fibre has an acceptance angle 26.80°. Calculate its numerical aperture. **(Ans: 0.4508)**
4. An optical fibre refractive indices of core and cladding are 1.53 and 1.42 respectively. Calculate its critical angle. **(Ans: 68.14°)**
5. Consider a fibre having a core of index 1.48, a cladding of index 1.46 and has a core diameter of 30 μm. Show that all rays making an angle less than 9.43° with the axis will propagate through the fibre.
6. A step-index fibre is made with a core of index 1.54, a cladding of index 1.50 and has a core diameter of 50 μm. It is operated at a wavelength of 1.3 μm. Find the V- number and the number of modes that the fibre will support. **(Ans: 42.15, 888)**
7. Using a step index fibre with n_1=1.48 and n_2=1.46 and the core radius a=30 μm. Calculate the number of total internal reflections that will occur on its propagation in a length of 1 km fibre.
8. A step-index fibre has a core refractive index of 1.44 and the cladding refractive index of 1.41. Find (i) the numerical aperture,(ii) the relative refractive index difference, and the acceptance angle. **(Ans: 0.292, 0.021, 33.96°)**
9. An optical fibre has a numerical aperture of 0.20 and a cladding refractive index of 1.59. Find the acceptance angle for the fibre in water which has a refractive index of 1.33. **(Ans: 8°39')**
10. Compute the cut-off parameter and the number of modes supported by a fibre which has a core refractive index of 1.54 and the cladding refractive index of 1.50. The radius of the core is 25 μm and operating wavelength is 1300 nm. **(Ans: 42.15, 888)**
11. Find the numerical aperture and acceptance angle of a fibre of core index 1.4 and $\Delta = 0.02$. **(Ans: 0.28, 32.52°)**

Made in United States
Orlando, FL
30 November 2024